More Praise for *Wedding of*

"*Wedding of the Waters* is a valuable history lesson for people—i.e., most of us—who have forgotten about the canal or never knew about it in the first place. Bernstein . . . gives the story of the canal's conception and construction all the drama it deserves. . . . It is almost impossible to imagine what the country would be like had it not been built. Peter Bernstein does it full justice."

—Jonathan Yardley, *Washington Post Book Review*

"Mr. Bernstein has opened a rich historical vein."

—Roger Lowenstein, *Wall Street Journal*

"Mr. Bernstein is at his best in showing how a colorful group of American politicians turned this dream into reality. . . . The book provides a splendid window into early American life, which was as raucously divided then as it is now by class and taste." —*The Economist*

"*Wedding of the Waters* is an important window into a vital and too often neglected period in the American past."

—Walter Russel Mead, *Foreign Affairs*

"A book that takes a sweeping, informed and often amusing look at the characters and visionaries who changed the history of America by digging a ditch across New York State. . . . [Bernstein] does a masterful job of placing the amazing achievement of the canal-builders in historic context, and in describing the Erie Canal's often-overlooked impact on not just local history but on the course of the nation and the development of a world economy that was changed by the Erie's sudden opening of the American heartland. . . . In this book, he reminds a world that had all but forgotten the Erie Canal that it was, in its day, the engineering wonder of the world, and that it fully deserved its fame."

—Mike Vogel, *Buffalo News*

"A riveting account of one of the most amazing technological achievements of all time. . . ." —Thomas J. Brady, *Philadelphia Inquirer*

"This is the epic narrative told in *Wedding of the Waters*, Peter L. Bernstein's engaging new history of the canal: the tale of how an engineering project changed America from colonial backwater to world power. . . . An exciting account of one of America's grandest civic projects. Bernstein gives us a colorful picture of a great undertaking by a country on the verge of greatness." —Arthur Vaughan, *New York Sun*

"Bernstein does a first-rate job of examining the social, political and economic impact of the canal both as a construction project and as a viable path linking the Atlantic seaboard with the American interior."
—*Publishers Weekly*

"An excellent overview of the struggles that went into making the Erie Canal a reality." —Josh Ozersky, *Chicago Tribune*

"Bernstein's economic analysis lucidly conveys the enormous impact of the Erie Canal and explains its crucial role in bringing about the industrial emergence of the young nation. Those interested in the confluence of history and economics will find Bernstein an especially valuable resource." —Chuck Leddy, *San Francisco Chronicle*

"One corner of the great American panorama enlarged to highlight starry-eyed visionaries, political machinations, indefatigable ingenuity, and cockeyed optimism." —*Kirkus Reviews*

"Engrossing. . . . Recounting the construction of the Erie Canal, the 363-mile ditch that connected the Western US territories to New York and, by extension, Europe and the world, the tale merits a place among the popular classics of American economic history."
—Stephen Schurr, *Financial Times*

"Bernstein's deft writing and deep understanding of global economic history makes *Wedding of the Waters* not just a convincing brief for the project's importance but a fun read as well."

—Nanette Byrnes, *BusinessWeek*

"The story behind the canal's construction is a narrative of political intrigue, bold vision and backbreaking work. And it all comes to life in Peter L. Bernstein's epic history of the canal. . . . Bernstein does a fabulous job of reminding us just how important Clinton's ditch was to the development of the city, the state and indeed the nation."

—Terry Golway, *New York Post*

"This is a wonderful story, an anodyne for our time. . . ."

—Joe Mysak, Bloomberg.com

"For every stripe of reader—history buff, news junkie, novel-devourer—*Wedding of the Waters* . . . will be a satisfying experience. . . . Gracefully written, rich in anecdote and fact, it's a book to be savored, talked about, and kept."

—Ann La Farge, *The Independent*

"The story, told by Peter L. Bernstein in his new book, *Wedding of the Waters*, isn't unknown, of course, but Bernstein freshens it up by setting it within the larger saga of a young nation striving to find its way."

—Wayne Curtis, *Preservation*

"In this latest work, Bernstein explains complicated subjects in digestible and delightful terms. What results is a brief history of the Erie Canal that's neither too exhaustive nor too shallow."

—Matthew Lubanko, *Hartford Courant*

"[Bernstein] explains the economic and financial aspects in a way that even I—usually baffled by these mysteries—could understand."

—Mark Dunkelman, *Providence Journal*

PREVIOUS BOOKS BY PETER L. BERNSTEIN

The Price of Prosperity

A Primer on Government Spending (with Robert Heilbroner)

A Primer on Money, Banking, and Gold

Economist on Wall Street

The Debt and the Deficit (with Robert Heilbroner)

Capital Ideas: The Improbable Origins of Modern Wall Street

Against the Gods: The Remarkable Story of Risk

The Power of Gold: The History of an Obsession

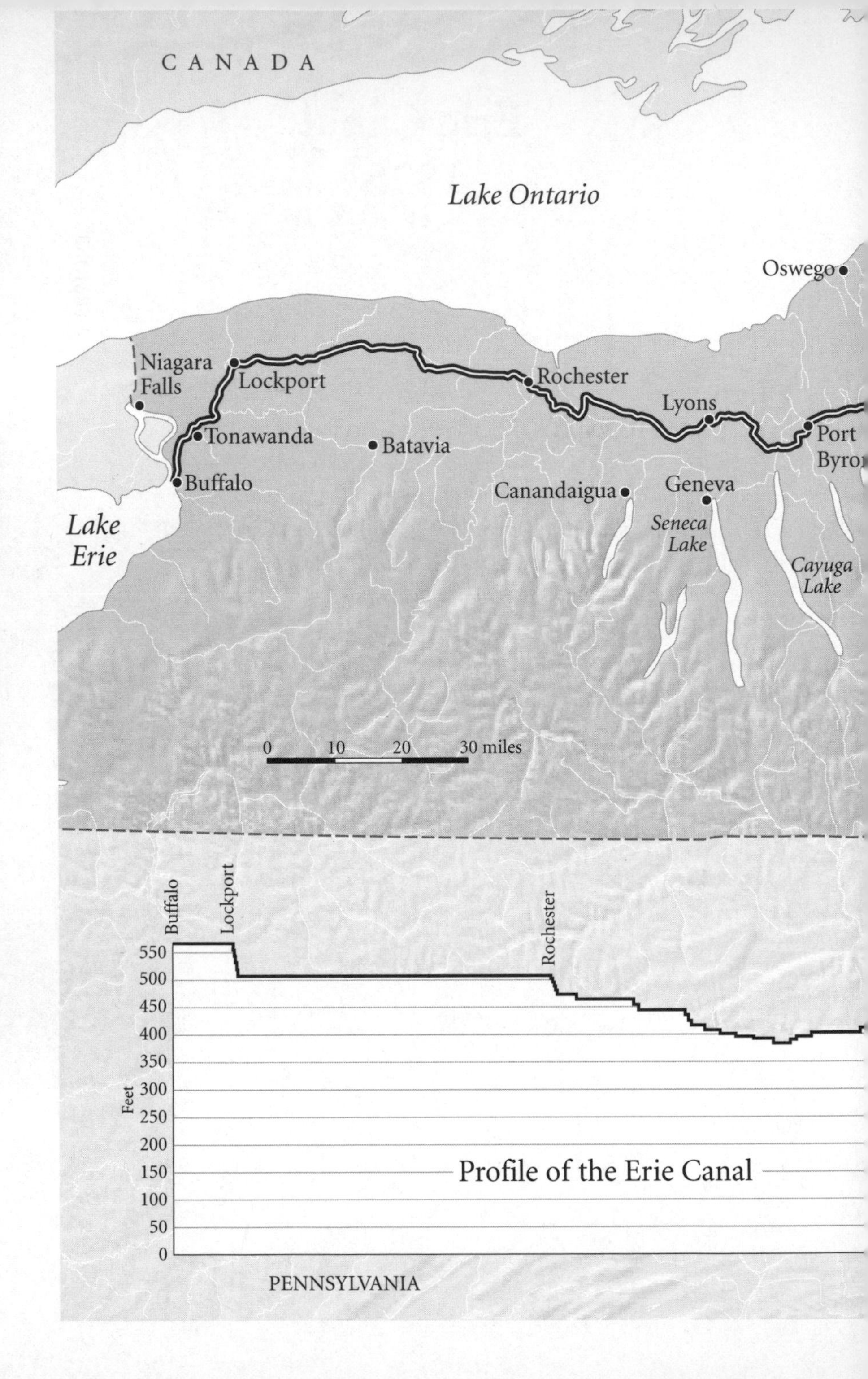
CANADA
Lake Ontario
Oswego
Niagara Falls
Lockport
Rochester
Lyons
Port Byron
Tonawanda
Batavia
Buffalo
Canandaigua
Geneva
Seneca Lake
Cayuga Lake
Lake Erie
0
10
20
30 miles
Buffalo
Lockport
Rochester
550
500
450
400
350
300
250
200
150
100
50
0
Feet
Profile of the Erie Canal
PENNSYLVANIA

Erie Canal
1825
eida Lake
Rome
Oneida
Utica
cuse
Mohawk
River
Little Falls
Canajoharie
Amsterdam
Schenectady
Troy
Albany
NEW YORK
VERMONT
MASSACHUSETTS
CONNECTICUT
Hudson River
Newburgh
Utica
Schenectady
Albany
NEW JERSEY
Map by Paul J. Pugliese

WEDDING *of the* WATERS

The Erie Canal and the Making of a Great Nation

Peter L. Bernstein

W. W. NORTON & COMPANY • NEW YORK • LONDON

In loving memory of my parents

Printed in the United States of America
First published as a Norton paperback 2006

Manufacturing by The Haddon Craftsmen, Inc.
Book design by Mary McDonnell
Production manager: Anna Oler

Library of Congress Cataloging-in-Publication Data
Bernstein, Peter L.
Wedding of the waters : the Erie Canal and the making of a great nation / Peter L. Bernstein.— 1st ed.
p. cm.
Summary: "The account of how the Erie Canal forever changed the course of American history"—Provided by publisher.
Includes bibliographical references and index.
ISBN 0-393-05233-8 (hardcover)
1. Erie Canal (N.Y.) 2. United States—Economic conditions. I. Title.
HE396.E6B47 2005
386'.48'09747—dc22

2004022792

ISBN 0-393-32795-7 pbk.

W. W. Norton & Company, Inc.
500 Fifth Avenue, New York, N.Y. 10110
www.wwnorton.com

W. W. Norton & Company Ltd.
Castle House, 75/76 Wells Street, London W1T 3QT

2 3 4 5 6 7 8 9 0

It certainly strikes the beholder with astonishment, to perceive what vast difficulties can be overcome by the pigmy arms of little mortal man, aided by science and directed by superior skill.

—*Henry Tudor, an English visitor to the Erie Canal in 1831*

CONTENTS

ACKNOWLEDGMENTS

Help and encouragement from others have played a more important role in this book than in my past endeavors. I was covering unfamiliar territory and encountered a research task that far exceeded my original expectations. My gratitude runs deep for all the generous assistance and advice I have received.

The most important person in this process was my wife, Barbara, who is also my business partner. I shall never find adequate words to express my boundless appreciation for her enthusiasm for the idea, for her tireless participation in searching for facts and concepts, for her hard-nosed editing and constructive criticism (she can tolerate four-letter words but those with three letters are very much at risk), and for her skilled management of all the other aspects of our busy lives while I was consumed with canals and Clintons.

Early on in my career of writing books, I learned how much difference a good editor can make. Drake McFeely, of W. W. Norton, has been a forceful and cooperative partner in this enterprise from the very beginning. Drake is positive, patient, and knowledgeable, but he is also firm and decisive. His illuminating influence shines throughout the book.

This book has been the fourth opportunity to have my wonderful friend Peter Dougherty as a prime support and creative critic. Peter has been my enthusiastic guide to what this great story was all about. His insights and understanding were essential in my discovery of the big ideas as well as in shaping and articulating them.

Brendan Curry of Norton has been an unusually helpful associate. Without his unremitting care and attention, the final stages of preparing the manuscript would have been laborious and attenuated instead of a pleasant and rewarding task.

I am deeply indebted to Trent Duffy, who painstakingly and skillfully reshaped a manuscript with many wrinkles, bulges, and gaps into a far smoother document than it was at the outset.

A special note of thanks is due Ken Galbraith. Ken inadvertently planted the idea for this book one day at his home in Vermont, when he explained to Barbara and me that many of Vermont's verdant forests replaced farmland deserted after 1825 by large numbers of their owners who rode the Erie Canal to the more fertile lands to the west.

I was fortunate in having the invaluable assistance of Professor Richard Sylla of the Stern School of Business at New York University in finding and interpreting many facts in the book. Dick also provided authoritative criticism and a meticulous reading of the manuscript.

Daniel Mazeau of SUNY Albany was a diligent, creative, and companionable research assistant. I could not have asked for better. His successor, Ralph Rataul, although only briefly on the job, carried on in the same spirit.

Morton Meyers of Buffalo was a great help in the research effort. I am happy to acknowledge his tireless efforts.

Professors Douglas Irwin of Dartmouth and Cheryl Schonhardt-Bailey of the London School of Economics made an important contribution to the conclusions and findings of Chapter 20.

The manuscript has benefited from the help provided by Philip Lord of the New York State Museum in Albany, who was generous in sharing his expert historical knowledge of New York and the Erie Canal.

Our business associate, Barbara Fotinatos, provided invaluable assistance and support to Barbara and me at all times in this long process.

Sam Rosborough was a valued and productive companion on our trip to the Erie Canal and the surrounding country in the spring of 2003.

I have one unorthodox acknowledgment: to the Internet. Having made only minimal use of the Internet in my previous books, I had no idea back then of the amazing quantity of information and analysis residing there. I am especially grateful to Bill Carr, who put online all of the material from Dr. David Hosack's invaluable memoir of De Witt Clinton and its extraordinary appendix of recollections by just about everyone who was in any way connected with the creation of the Erie Canal. That was just the beginning. A quick glance at the endnotes will reveal the critical importance of the Internet as a research source.

Many others were helpful in a wide variety of ways. In particular, I want to thank Keith Ambachtsheer, Peter Brodsky, Tom Green, Ned Goldwasser, Alan Greenspan, Judith Heller, Representative Carolyn Maloney and her staff, Brad Seymour, Michelle Smith, Lawrence Summers, Richard Sutch, Richard Tedlow, Erin Walsh, and Julian Zelizer.

WEDDING *of the* WATERS

INTRODUCTION

"DOES IT NOT SEEM LIKE MAGIC?"

We live today in a time of widespread concern and fascination with technological innovation, the system of networking, and the influence of both on globalization in the twenty-first century. We also live in a time when the United States faces bewildering questions about its role in the world and, indeed, the vision of its long-term future. This acceleration in change has affected our political system, our society, the economy, and the world of finance in ways we are only beginning to comprehend.

This book draws upon the story of the Erie Canal to illuminate the turbulent and exciting present with a great but unfamiliar history of how a revolutionary technological network molded the triumph of the United States as a continental power and as a giant in the world economy. The Erie Canal was the child of many dreamers and a host of surveyors, engineers, and politicians, most of whom had never seen a canal before and few of whom had any experience in designing, building, or operating a canal. But the heroes of this story had the foresight to change the face of the earth, not only literally but in a much more fundamental sense. They understood that the process they launched would alter every aspect of how people lived their lives.

When the canal was completed in October 1825 and Governor De Witt Clinton could celebrate the Wedding of the Waters by pouring a keg of water from Lake Erie into the Atlantic Ocean, he opened an uninterrupted navigable waterway through the imposing barrier of the mountain range extending from Maine all the way down to Georgia. The end result would lead to an historic explosion of commerce, ideas, and technological change. By bringing the interior to the seas and the seas into the interior, the Erie Canal would shape a great nation, knit the sinews of the Industrial Revolution, propel globalization—extending America's networks outside our own borders—and revolutionize the production and supply of food for the entire world. That was by no means all. In time, this skinny ditch in upstate New York would demonstrate that trade and commerce are the keys to the expansion of prosperity and freedom itself.

* * *

The notion of a canal connecting the eastern seaboard to the west had been an active topic of discussion for more than twenty years before construction actually began in 1817. The Appalachian mountain range posed a formidable barricade between the narrow line of states touching the Atlantic Ocean and the almost boundless lands on the other side of the mountains. While rivers often lead the people on opposite sides of their banks to join in forming one community, populations divided by mountains tend to become separate nations unless some easy means of communication exists between the two.

George Washington was keenly aware of this risk. Even before the Revolution, in 1775, he had expressed his concerns about the peril of losing the lands on the western side of the Appalachian Mountains to either France or Canada, or both, unless the mountain barrier could be pierced—and soon. The pioneers moving west had little allegiance to the lands they left behind. If nothing were done, the young United States would be left squeezed between the mountains and the sea, a constricted minor-league nation compared with the growth and power developing on the other side of the mountains.

Ink on the peace treaty declaring American independence was barely dry before Washington was organizing the Patowmack Company to convert the Potomac River into a canal running from the seacoast at Alexandria, Virginia, all the way up to the mountains. Washington's canal was an engineering achievement but a financial disaster. It was still under construction when he died in 1799 and fell into bankruptcy before work on the Erie Canal had even begun.

The geography for crossing the mountains was much more favorable in New York than in Virginia, but New Yorkers appeared to be in less of a hurry to capitalize on what nature had bestowed upon them. There was a succession of frustrating efforts over a period of some twenty-five years before they would finally break ground for the Erie Canal in 1817. The visionaries and supporters were determined to pursue their dream. Their opponents were stubborn in their skepticism about the feasibility of such a gigantic engineering venture, and many were also frightened by the prospect of spending so many millions of dollars of the state's money with no assurance that the canal would actually pay its way.

To Americans in the early days of the Republic—in what was then, in today's parlance, an "emerging economy" in which 93 percent of the population still lived on farms—the very idea of connecting the east to the west by means of a gigantic artificial waterway appeared as fantastic as sending a rocket to the moon. In January 1809, Thomas Jefferson judged it to be "little short of madness." Without the gritty determination of a small group of men convinced of the prospect of a great nation, an unquenchable enthusiasm to make it reality, and a keen sense of how to deploy power, the Erie Canal would not have been built and the western territories would in all likelihood have broken away.

* * *

New York's first attempt to develop an effective passage through the mountains was, like Washington's Patowmack Company, a privately financed venture; it was established in 1792 under the leadership of General Philip Schuyler, a hero of the Revolutionary War, and a merchant named Elkanah Watson who had all the right political connections. Rather than build a canal, the designers of this ill-fated effort attempted to transform the Mohawk River's many rapids, falls, and shallow stretches into a navigable waterway over about a hundred miles from Albany on the Hudson to the western interior of New York State. Engineering this project was difficult enough, but the lack of sufficient revenue to cover the enormous maintenance expenses turned out to be an even greater hurdle. Like the Patowmack Company—and for many of the same reasons—this endeavor ended in financial failure.

A few New Yorkers kept the faith and continued the struggle. There were those, such as the irrepressible optimist Gouverneur Morris, who proclaimed with remarkable clarity both the possibilities and huge benefits that a canal would bestow on the state and nation. Two members of the state legislature began to press for a canal around 1805, but most people tended to agree with Thomas Jefferson that huge financial costs and the scale and audacity of the engineering were insuperable obstacles.

Finally, in 1810, a small cadre of enthusiasts from the state legislature enlisted the support of De Witt Clinton, the most powerful politician in the state at that moment. A serious intellectual and amateur scientist, Clinton was an insider's insider in the world of politics with a well-honed instinct for the uses of power. He had been the secretary for his uncle George Clinton, the nine-term governor of New York State who was then vice president of the United States. He had served as a U.S. senator, had been mayor of New York City for six years, and was now a state senator. He was about to become lieutenant governor.

Clinton's decision to participate was the pivotal moment in this story. He was the one individual with the political skill and intellectual authority to turn Morris's bold vision into a reality. Once he considered the possibilities, Clinton was willing to put his entire future on the line for the canal, without reservation and without fear of the political turmoil looming before him. Despite his high-handed mannerisms and unquenchable ambitions for high office, which tended to spawn enemies among politicians even in his own party, Clinton's firm leadership of a reluctant electorate, his extraordinary eloquence, and his brilliant analytical capabilities brought the boundless benefits of the canal to New York State and, ultimately, to the United States of America. There might never have been an Erie Canal without his unwavering support and readiness to risk his career on the canal's success.

The political battles and intrigues fought out in New York State over the fate of the Erie Canal were as fierce as any in its history. Politicians high and low, and from all parties, engaged in this struggle tirelessly and with great zest. The controversies engaged not only eminent leaders like De Witt Clinton and Martin Van Buren—both of whom made no effort to disguise their presidential aspirations—but the bare-knuckled forces of Tammany Hall as well. On more than one occasion, victory seemed within grasp only to be dashed by the resiliency of the opposition and the incredulity of the timid. In many ways, it was a lot easier for the engineers to improvise this stupendous technological achievement than it was for De Witt Clinton and his allies to subdue their political opponents.

And as the time finally arrived when, as one protagonist put it, "it was impossible for stupidity itself [not] to stretch their opaque minds from Erie to the ocean," the former opponents of the canal turned themselves right around and claimed to be the true guardians of its future. Now the canal developed into a kind of religious totem, perceived as above and beyond the rambunctious political arena.

* * *

When completed in 1825, the Erie Canal was the marvel of its age. As one of many tourists who came to see this technological triumph summed it up, "The truth is the canal is in everybody's mouth."[1] Yet this completely man-made waterway was dug and hacked through dense forests and built over rivers and valleys with nothing more than bare hands, shovels and axes, mules, explosive powder, and crude but ingenious inventions to pull down trees and yank up their stumps.

The canal stretched 363 miles from Buffalo on Lake Erie, part of the Great Lakes system (referred to by Clinton as "our Mediterranean seas"), to Albany on the Hudson River. From Albany, it was 150 miles down the Hudson to the vast harbor of New York City and its well-developed port facilities. Designed by a small group of surveyors with no civil engineering experience but a great talent for improvisation and innovation, the canal's eighty-three locks enabled boats to travel a total of 675 feet up and down from one end to the other, over hill and dale, and over eighteen aqueducts. One aqueduct more than three city blocks long carried the canal so far above the raging river below that the Marquis de Lafayette would describe it in wonder as "an aerial route."[2] The canal was also crossed by hundreds of small bridges connecting the lands it sliced through on its path between Albany and Buffalo.

Once it was opened, the traffic on the canal expanded rapidly. A cornucopia of western agricultural products moved eastward toward the Atlantic while manufactured goods and, increasingly, immigrants and other pioneers traveled westward to the limitless undeveloped lands lying in wait. By replacing the rough roads, the Erie Canal slashed transport costs so radically that a shipment of flour could have traveled 2750 miles before the cost would exceed 50 percent of its ultimate proceeds.[3] By road, the equivalent distance was only 130 miles. As a result, the tonnage of grain reaching Buffalo from the west in the mid-1840s was ten times what it had been just a decade earlier.[4] The Erie Canal's agricultural cargoes soon found markets throughout Europe. In the process, the canal would influence such mighty decisions as the repeal of tariffs on food in Great Britain in 1846, freeing European labor for work in the factories and reducing costs of production throughout the European as well as the American economy.

The prosperity created by the canal earned New York the proud title of the Empire State* and made New York City the greatest city in the nation and, some would argue, in the world. Indeed, the Erie Canal would have a compelling influence on the Industrial Revolution, as well as the perceptions, and the realities, of economic power. The canal counties of New York were early and sustained leaders in the number of patents issued in the United States each year. As late as 1852, freight tonnage carried on an enlarged Erie Canal was still thirteen times the traffic carried on all the railroads in New York State.[5] When tolls were abolished in 1882, the canal was serving over twenty million people and had produced revenues of $121 million since 1825, more than quadruple its operating costs.[6] And it was still going strong.

The financing of the Erie Canal was as original and brilliantly successful as its technological features. When the U.S. government in Washington finally rejected sponsorship of the canal in 1816—after repeated promises to contribute—New York State proceeded to finance the canal by selling its own bonds to the public at large and, with rising success, to financial markets abroad. The flow of revenues collected on the canal was so far in excess of operating expenses that the state was able to repay these bonds well ahead of the schedule.

Most important, Americans perceived the canal as an expression of faith in the potentials of a free society, a message of hope for a great young nation on the move. All of the leaders of the campaign for the canal—De Witt Clinton, Gouverneur Morris, Robert Fulton, as well as others less famous—shared a cloudless view of how the future would unfold if the commercial connection between east and west could be secured. Their foresight was remarkable: almost every feature of their wonderful dream came true.

"Does it not seem like magic?" Gouverneur Morris had asked a

*The derivation of this expression is lost in the mists of the past. It appears to have been in general use by 1820, although there are scattered mentions of it earlier than that. For more detail, see Milton Klein, *The Empire State*, p. xix.

friend in December 1800, seventeen years before the first ground would be moved for construction of the canal. At the culmination of a trip to look over his own holdings of real estate across the raw backcountry of New York State, Morris went on:

> Shall I lead your astonishment to the verge of incredulity? I will: know then, that one-tenth of the expense born by Britain in the last campaign would enable ships to sail from London through Hudson's river into Lake Erie. Hundreds of ships will in no distant period bound the billows of those inland seas. . . . As yet, my friend, we only crawl about on the outer shell of our country. The interior excels the part we inhabit in soil, in climate, in everything. The proudest empire in Europe is but a bauble compared to what America *will* be and *must* be, in the course of two centuries, perhaps of one.[7]

PART I

The Visionaries

CHAPTER 1

SMOOTH SAILING

On October 26, 1825, a line of tiny boats departed from Buffalo, New York, sailing eastward from the shores of Lake Erie to the Hudson River at Albany. But there is no river running from Lake Erie to the Hudson. Instead, the boats traveled on an artificial waterway for 363 miles, propelled forward only by horses plodding along the shore.

This curious event marked the start of the first complete voyage on the Erie Canal, with Governor De Witt Clinton aboard the lead boat. Nine days later, at 7:00 a.m. on the morning of November 4, the boats arrived at Albany. Now steamboats replaced the horses, pulling the four leading canal boats as they turned southward down the Hudson River toward New York City. Wild cheers from the crowds along the shores greeted the little flotilla as it glided past bonfires and cannon salutes along the river's banks all the way to New York harbor. When De Witt Clinton stepped to the prow of the escorting warship *Washington* and slowly poured the keg of Lake Erie water into the Atlantic, the ceremony called "the Wedding of the Waters" was complete.

* * *

It is fair to ask what can be so compelling about canals. Why was there so much excitement over the opening of the Erie Canal in 1825?

Transportation has been a formidable challenge for most of human history. Villages just twenty miles apart once seemed far away from each other. Only urgent business prompted high-ranking nobles and priests to undertake the perils and discomforts of travel over longer distances. Chaucer reminds us that the great pilgrimages of the Middle Ages were possible only because of safety in numbers. But how many people, even among the most pious, could leave their fields to toil along the roads on a voyage lasting months and occasionally over a year?

Today, people travel constantly. They go to work, to shop, to restaurants, on business, to visit friends and family, to sightsee. Their trips take them over dirt roads and paved roads, on rivers, across oceans, and into the skies. They go around the block or thousands of miles in cars, taxis, subways, buses, ferries, trains, airplanes, and ships, without a moment's hesitation, and barely a thought about the miracles transport has created for them. Meanwhile, billions of tons of cargo move along the same routes.

This extraordinary system helps to define what we mean by "modern times." Today's global complex of routes is the glue that binds the world together, a vast network transporting global commerce and linking individuals, corporations, and governments. Transportation can compress into a matter of hours the great spaces between Omaha and Beijing, between Avignon and Johannesburg, or between Buenos Aires and Sydney with the same ease that a daily commuter rides from the suburbs to work downtown. Telephones and computers may be able to transmit words, data, pictures, and messages instantly, but neither wire nor wireless can move a human body, a ton of steel, or even a single microchip.

The Erie Canal, and the waters it wed, played a key role in leading the modern world toward these miraculous systems. Despite the brilliance of its design and execution, the Erie was just one step in the ancient art of defying nature to shape the flows of water to the needs of man. In the beginning, long before that, people knew only how to walk on their two legs from one point to another. Many things had to be learned before humans could transport themselves on a canal.

* * *

The premier innovation in transportation was the invention of the wheel, and the next most important was placing a sail on a ship in order to capture the energy of the winds to move great distances. Neither was sufficient to bring the world together in the way we know it today. Although wheels moved people on land and sailing ships transported them across the seas, you cannot sail across land, and for most of history the wheel guaranteed neither speed nor comfort. Most people moved from place to place down wretched dirt paths or along muddy and rocky roadways. They traveled on foot, often with heavy packs on their backs, or in rough wooden carts jerked along by plodding beasts of burden. Although upper classes in some parts of the world could travel on horseback, there was nothing comfortable or fast about that either.

With travel over land so disagreeable, precarious, and slow, water and boats were the vehicles of choice for trips of more than a few miles, especially for bulky items like grain and timber that constituted most of economic output in ancient times. As a consequence, the leading towns and cities tended to rise on the banks of rivers or at the ocean ports where the human and physical cargo arrived at the sea from the inland waterways. A piece of old Swedish doggerel tells of the foolish man:

> In his journeys he noticed,
> How well the providence of God has disposed,
> Who put rivers in all the places
> Where big cities happened to be.[1]

Great natural waterways go far to explain the rapid pace of economic development and economic power in North America. As one observer pointed out as far back as 1804, when the United States was still just thirteen states lined up along the Atlantic coast plus Vermont, Tennessee, Kentucky, and Ohio away from the coast, "The facility of navigation renders the communication between the ports of Georgia and New Hampshire infinitely more expeditious and practicable than between those of Provence and Picardy in France, Cornwall and Caithness in Great Britain, or Gallicia [*sic*] and Catalonia, in Spain."[2]

* * *

While you can travel in either direction on a road, river water flows in only one direction, dutifully responding to where gravity and the shape of the earth tell it to go. As a result, the places where people built their cities and the water routes they followed were hostage to the natural forces forming the paths of rivers and the proportions of lakes. Movement downstream on many rivers was little problem, but traveling upstream against the current, and transporting heavy cargo in particular, was often an insurmountable obstacle. Without some means of breaking free from the constraints imposed by nature, water would be only an occasional alternative to the trudging animals and the cruel swellings and cracks on the surfaces of the roads.

About four thousand years ago, a few enterprising people discovered an audacious method to liberate travel and shipping from nature's grip: they began digging extended water-filled ditches between one river and another, around river obstacles like waterfalls and rocky bottoms, or even across land without regard to rivers. These man-made waterways came to be known as canals, a word derived from the Latin *canalis*, meaning "pipe" or "channel."

Canals would change everything. By employing engineering skills and the sweat of human labor to overcome the routes so arbitrarily decreed by nature, people were able to enjoy a far wider range of choices as to where they would live, how they would transport the fruits of their daily work, and how they would travel from place to place. Canals were the origin of networking in the fullest sense in which we think of it in modern times.

Nevertheless, nature interferes even with canals. The whole purpose of a canal is to hold the water in the ditch essentially motionless so that boats can move with equal ease in either direction. But the earth is seldom flat: hills and valleys intervene at nearly every step of the way along the planned route.

The earliest known method to confront this problem was a sequence of dams across the waterway—a crude form of what came to be known as locks. When a boat approached in the downstream direction, a gate in the dam was raised. The boat would then shoot through the dam with the full force of the water behind it. These locks, for good reason, were called flash locks. But the flash was a lot less of a thrill for boats traveling uphill. The gate in the dam was still opened as the uphill boats approached, but then gangs of men or teams of animals had to pull them forward against the current. When the water pressure was too great, the laborers had to carry the boats up an inclined plane or over land around the dam. This crude procedure prevailed everywhere until about AD 1000.[3]

Even with these rough-and-ready methods, the Babylonians dug a canal around 2200 BC to connect the Tigris and Euphrates rivers. At about the same time, Egypt was improving navigation along the Nile, including building a canal to bypass the cataract at Aswan. In the twelfth century BC, the Egyptians would build a canal connecting the Nile to the Red Sea. A trip on these canals was a joy after the rough and tumble of travel overland. The Book of the Dead described sailing "where the reeds are continually bending in the heavenly wind" as one of the pleasures of Paradise.[4]

Both the Babylonian and the Egyptian canals were built over relatively flat and sandy territory, but the Chinese were more ambitious.[5] Their canal system dates back as far as 600 BC, when construction began on what was to be the Grand Canal, which extended over a thousand miles from Tianjan near Beijing in the north to Hangzhou near Shanghai, making it the world's longest man-made waterway. With no construction machinery of any kind, the millions of peasants forced to labor on the canal scooped up the earth with nothing but shovels and often with their bare hands; about half of these poor creatures died at their backbreaking work. Some of the work simply improved navigation on rivers, but significant parts formed a sequence of canals fed primarily from nearby lakes along the route.

The Chinese toiled and drudged over this prodigious project in fits and starts over the centuries. Most of the time, the fits outnumbered the starts. More than a thousand years would pass before the separate sections were finally linked up into one uninterrupted canal during a spurt of frenetic activity from 605 to 610 AD.

The canal brought the great expanse of China to life. As the major rivers in China run east and west, the north-south route of the Grand Canal connected waterways such as the Yangtze and the Yellow rivers. This achievement transformed the economic patterns of centuries by shifting the center of food production from the wheat- and millet-producing regions of the north to the rice fields of the south and also formed the basis for a unified economy and the development of Chinese power in Asia.[6]

The Romans built a fifteen-mile canal in the first century BC, running across the Pontine Marshes parallel to the Appian Way to carry passengers when floods made the road impassable. Nero made two efforts to build canals—one between Rome and Naples and the other across the Isthmus of Corinth in Greece in AD 65—but neither succeeded. (The latter was rectified only in 1893.) About fifty years later, the Romans in Britain did succeed in finishing the Carr and Foss Dykes from Peterborough to Lincoln. The Dutch built canals with locks as early as 1065, the Italians followed suit a hundred years after that, and more elaborate lock systems were common by 1500. In 1601, Samuel de Champlain, one of the great explorers in North America, published his *Brief Narrative* describing his voyage to the West Indies, in which he proposed a canal across Panama, then known as Darien, to connect the Atlantic to the Pacific.

By that time, canal engineers had achieved major advances in the design of locks to raise and lower boats up and down hills along the route of a canal. Pioneering studies on lock design derived from Leonardo da Vinci (1452–1519), who was obsessed with the movement of water. Among his countless drawings of water in motion, Leonardo left plans to reroute the Arno and to build canals on the Loire. He is also credited with the design of eighteen locks, climbing eighty feet, on the Duke of Milan's Bereguardo Canal, built in 1485 to connect the Pavia area to Milan.[7]

A typical lock, as designed by Leonardo and still at work today, consists of a chamber along the path of the canal with a pair of gates at both the upstream and the downstream ends. Unlike the flat dams of flash locks, Leonardo's special innovation was to shape these gates into a V pointing upstream when closed. The pressure of the current pushes against the V, keeping the gate shut tight so that no water can leak into or out of the chamber. While the gates in the dams in flash locks let the water in and out by moving up and down, Leonardo's gates swing open and closed. Sluices in the gates allow the water to flow through down to the next level. When hills are long and steep, one lock frequently leads a boat directly into the next lock above or below it, so that the sequence forms a stairway of locks, usually called a flight. If you tried to drink all the water in a lock built to Leonardo's specifications that could lift a boat by about ten feet, you would have to consume 5.6 million cups of water.[8]

A boat approaching the canal on the upstream side would be heading downhill. After it enters the chamber and the gates are closed behind, the water in the lock empties out through the sluices in the gates at the downstream end until the level of the water in the lock drops to the level of the waterway on the far side of those downstream gates. At that point, the gates open up and the boat proceeds on its way. Thus, the lock functions as a kind of elevator carrying the boat downhill.

A boat heading uphill enters through the lower gates, which are then closed behind it while the sluices in the gates at the upper end are opened, filling the lock to the level of the waterway on the upstream side. Once again, the gates are opened and the boat proceeds on its way at the top of the hill.

* * *

One of the most spectacular European canals was the Canal du Midi, built from 1666 to 1681 under the reign and with the blessings of King Louis XIV. This great project carved a water route right across southern France from the Atlantic to the Mediterranean, an idea dating as far back as the Romans and Charlemagne but never before brought to full fruition.[9] By allowing ships to avoid the tedious and often dangerous Atlantic voyage around Spain and back into the Mediterranean through the Straits of Gibraltar, the Canal du Midi revolutionized the economy and civilization of southwestern France. A hundred years later, Voltaire would describe the canal as "*le monument le plus glorieux.*"[10]

The Languedoc, the region served by the canal, is blessed with fertile soils that produce large quantities of corn, fruit, olives, saffron, mulberries, and berries whose colors were in great demand for dyes. But in the mid-seventeenth century the Languedoc was also a hotbed for brigands in a country where crumbling roads and bridges made travel painfully slow, leaving travelers and their cargoes as easy targets. If moving the precious crops out of the area was perilous by land, prowling pirates and stormy weather made sailing from Bordeaux around Spain to enter the Mediterranean even more hazardous. As a result, the whole southwestern tier of France stagnated for want of proper transportation facilities, a backward land falling ever further behind the prosperous provinces to the north and east, where large rivers like the Rhône, the Seine, the Loire, and the Meuse made transport relatively easy.

The Canal du Midi changed all that. A ship could enter the canal at Bordeaux, sail eastward on smooth waters for 150 miles across the midsection of France, pass through major urban centers such as Toulouse and the walled medieval city of Carcassonne, and exit into the northwest corner of the Mediterranean at Sète near Montpellier.[11] The voyage was spectacular, climbing to an altitude of 620 feet above sea level and back down again via 328 structures, including bridges, the first canal tunnel in Europe (500 yards in length), dams to help feed water into the canal, and 103 locks—20 more than the Erie Canal would require 150 years later. Unlike most of the other canals built up to that time, the long section of the Canal du Midi from Toulouse to the Mediterranean exit was an artificial waterway, enabling the builders to avoid the difficult problem of making unreliable natural waterways navigable. The designers of the Erie Canal would make the same choice.

Unlike rivers, artificial waterways need a constant supply of water throughout their entire length, even when the canal is more than 400 feet above sea level. The French canal designers solved this problem with the construction of the world's first known artificial reservoir for canals, fed from mountain streams and held in place by a massive dam 2300 feet long, 100 feet above the riverbed at its maximum, and up to 400 feet thick at its base.[12] This project required the labor of squads of women conscripted from the area, carrying the heavy earth up the hills in baskets, at a wage of a penny a basket. Two hundred and forty years later, in 1906, the Americans would duplicate this work in designing Gatun Lake on the Panama Canal.

Leaky earthen walls are among the biggest problems canal builders have to confront. The leaks can run either way—seeping into the canal from underground springs and neighboring rivers or draining water out of the canal into the surrounding earth. The seventeenth-century builders of the Canal du Midi overcame this difficulty with another remarkable innovation: they "puddled" the claylike soil used for the canal walls—a process of kneading and massaging the earth by manual labor over and over and over, and then over and over and over some more, until sufficient oxygen has been blended into the soil to make it impermeable. The sides of the Erie Canal, among many others, are sustained in this manner.

The Canal du Midi was the brainchild of Baron Pierre-Paul de Riquet de Bonrepos, a self-made man who amassed a fortune as a tax collector under Louis XIV. Colbert, the king's finance minister, was sufficiently impressed to arrange for Riquet to obtain partial financing through loans from the province. Riquet was consumed with his mission and undeterred by countless obstacles of political approvals, resistance from locals who saw all change as dangerous, and the seemingly bottomless financial requirements of the undertaking. Although Riquet enjoyed a direct line of contact to the king through Colbert, wars and other foreign distractions kept diverting the king's attention from providing the necessary money to complete the canal. That did not stop Riquet. He was so obsessed by his creation that he consumed his entire fortune paying for it. Without his personal finances, which he never recouped in full, the Canal du Midi might have ended up as nothing more than an unfinished and empty ditch.

* * *

The most innovative and influential man-made waterway of the eighteenth century was the Bridgewater Canal, an English creation constructed to connect a coal mine to Manchester, less than ten miles away—a dramatic contrast to France's "*monument le plus glorieux*." This midget waterway, with an aqueduct that carried the canal across the River Irwell at the level of the treetops, was a technological sensation that transformed the face and future of Britain with a fever of canal construction, greatly hastening the pace of the Industrial Revolution.[13]

The Bridgewater Canal was built from 1759 to 1761, a business speculation conceived and financed by Francis Egerton, who carried the eponymous title of the Duke of Bridgewater. He had been a sickly child but grew up into a tall man with a head just a bit too small for his height. He had a conventional education for a young aristocrat, including a grand tour of the Continent, during which he paid a visit to the Canal du Midi. It left a lasting impression. Egerton returned to England from France for a life of gaiety, drinking, gambling, and betting on the races—until he met Bridget, Countess of Mayo, and fell madly in love with her. But Bridget had other plans: she deserted him for the future Duke of Argyll, an older man with a higher ranking in the nobility and reputedly more skilled at the arts of love.

If Egerton's beloved had been more faithful to him, it is quite possible that the blossoming of the Industrial Revolution in Britain and the economic power that financed the British Empire would have arrived much later. Luckily for Britain, he turned to the project of a canal to distract him from his sorrows and mend his broken heart. This therapy changed the course of history.

For centuries, the English had been so focused on the seas around them that they paid little attention to the development of internal means of transportation. England has no great rivers like the Rhône, the Rhine, the Yangtze, the Nile, or the Mississippi, all of which carry heavy loads for inland transport as well as nourishing great seaports. This concentration on the oceans made little difference as long as almost all the major cities of England were on the seacoasts and interconnections within the interior were unnecessary. What little overland movement took place could be left to the creaking wagons and trudging packhorses. With the arrival of the Industrial Revolution, however, economic activity shifted inland toward the coal mines, creating a need for overland transport of massive quantities of coal as well as iron ore and machinery.

The young Duke of Bridgewater was blessed with a rich mine of coal on his properties, which he had inherited from his father, Scroop. And what a bequest! A coal mine in those days was as magnificent an asset as a gold mine. The greatest technological innovation of the eighteenth century—and among the greatest of all time—was steam: you could work miracles merely by boiling some water. But in the 1700s there was no way to generate steam in sufficient quantities without coal, and the middle of Britain sat on what appeared to be an inexhaustible supply of coal, some cropping up on the surface, but most of it buried underground in mines.

Francis Egerton's insight was to recognize that he could cut the price of coal in Manchester in half if he could ship the output of his mines over the rugged countryside by water instead of by road. In view of the rapidly expanding demand for coal to drive Manchester's burgeoning textile mills, the precious output of his mine at rock-bottom prices would grab a dominant share of the coal market. Success would multiply his fortunes. Nothing would serve better toward repairing his broken heart.

* * *

The project was risky and complicated, as the best path for a canal from the mine to the city was by no means obvious. The most direct route was hilly and almost entirely across dry land rather than over the local streams. But the duke was determined to prove that the shortest distance between two points was a straight line. He would build right across the countryside from the mine to Manchester, taking advantage of the local streams to furnish water for his boats to travel along his canal.

This revolutionary solution to the problem would lead many to call the duke a madman and to make laughingstocks out of John Gilbert, his agent and main assistant, and James Brindley, his chief engineer. Worse, the revolutionary enterprise would nearly lead the duke into bankruptcy. At one point, he could not find a bank in Liverpool or Manchester to cash a £500 check for him. On another occasion, Brindley had to restore his spirits by crying, "Don't mind, Duke, don't be cast down; we're sure to succeed after all."[14]

In the end, the great adventure would turn out to be a marvel of engineering. The canal's most spectacular feature, built with the sweat of four hundred men, was the aqueduct supported by three great arches to carry the canal across the River Irwell, about halfway between the coal mines and Manchester. People stared openmouthedly as boats carrying as much as twelve tons of coal approached the aqueduct by moving majestically forward on a long embankment, pierced by a tunnel for the main road below, and then proceeded to float for two hundred yards over the three arches, at a height of thirty-eight feet above the river, to the other side. As one contemporary enthusiast put it,

> Seen and acknowledged by astonished crowds
> From underground emerging to the clouds;
> Vessels o'er vessels, water under water.
> Bridgewater triumphs, art has conquered nature![15]

The poet's allusion to "from underground" refers to another marvel conceived by the duke, Gilbert, and Brindley. We have a description of this achievement from Elkanah Watson, who would later play an important role in the creation of the Erie Canal, inspired in part by his visit to the Bridgewater Canal in the 1770s. Watson caught the miracle of canal building in just a few words: "Not content with skimming along the surface, with traversing valleys and crossing rivers by their artificial navigation, they decided to plunge into the very bowels of a mountain, in pursuit of coal."[16]* In order to facilitate the delivery of coal to the canal, they excavated an intricate and interconnecting series of underground waterways covering forty-six miles within the coal mine itself, some of the many branches reaching as deep as four miles toward the mine shafts. That way, the coal could move right out on the waters of the canal, inexpensively, rapidly, and smoothly, instead

*The quotation is from Watson's *Memoirs*, published in 1856. Pages 164–66 of this volume contain an extended and vividly detailed description of all aspects of the Bridgewater Canal in the 1770s.

of being hauled by human labor to some place for shipment to its ultimate customers.

All the duke's fondest dreams came true. Coal prices fell in Manchester to two shillings sixpence a ton, which was a deeper cut even than the 50 percent he had guaranteed. As money poured into his pockets, Bridgewater became one of the richest men in Britain. The skeptics who had called him and his colleagues madmen were silenced.

In addition, James Brindley was now recognized as the foremost engineer of his age—an extraordinary achievement for a man who could barely read, seldom put pen to paper, and had started his career as plowboy before becoming a maintenance man on the water mills that powered what little machinery was then in use. Brindley would go on to design what was called the Grand Cross, a system of east-west-north-south canals reaching south of Manchester toward the rapidly growing potteries in the Birmingham area. In time, the Grand Cross made it possible to travel by water from London to Scotland and from East Anglia to Wales. By 1780, canal building had become a frenzy across all of England—including a western extension of the original ten-mile Bridgewater Canal to the great port of Liverpool, giving Manchester a direct connection by water to the Atlantic Ocean. Before they were done, the British had built more than a thousand miles of canal routes in an intricate network covering the entire Midlands and the center of their future industrial power. The excitement about canals prompted a mania and bubble on the stock exchange similar to the early days of railroad development, with overbuilding and high-visibility bankruptcies. Even so, like later forms of irrational exuberance in more modern times, the longer-term rewards from revolutionary technological innovation would remain.

Through it all, Brindley remained the country hick, no matter how much success he achieved. At one point, when he went to London to testify before Parliament on the duke's behalf, Gilbert took him to the Drury Lane to see David Garrick in Shakespeare's *Richard III*. Brindley had never been near a theater. Brindley was so shocked by the betrayals, murders, and seductions on the stage that he was laid low and could not leave his bed for several days. As soon as he could pull himself together, he set off to repent at morning service at the nearest church.[17]

The duke's dreams for his canal came true. His waterway moved coal from the mine to Manchester smoothly, rapidly, and with a huge reduction in its price. Later, his canal's significance reached far beyond the ditch and the aqueduct as the duke connected Manchester by canal to the sea, stimulating rapid growth in the greatly diversified traffic flow.* That was just fine as far as the duke was concerned. The low energy costs made possible by Bridgewater transformed Manchester into the textile manufacturing capital of the world.

* * *

All this canal building, launched by the great innovations at Bridgewater, was the prelude to the modern world economy, a system people could never even have dreamed of without the spectacular reduction in the costs of transportation provided by the canals, and the manner in which they linked inland producers with one another and with the great ports on the coasts. Although the canals made an enormous contribution to the development of the Industrial Revolution in Britain, the new economy's center of gravity would in time turn out to lie several thousand miles westward, at the gateway of New York harbor. It was there that the technology and ingenuity of the great canal makers of the past would connect with the vision and determination of a small group of men to raise commerce and democracy on their vast continent to a level unimaginable before then.

*Manchester, as probably the world's largest importer of raw cotton, has always been a strongly pro-American community. A statue of Abraham Lincoln stands in one of the city parks.

CHAPTER 2

HUDSON'S WRONG TURN

On September 10, 1609, Henry Hudson, famed English explorer in the spirit of the Elizabethans, steered his ship, the *Half Moon*, into New York harbor in search of the water route that would take him westward into the Pacific Ocean.* Like others before him, Hudson was convinced of the existence of the Northwest Passage, a means of traveling directly from Europe through North America to Asia without having to cross any land.

New York harbor had first been explored in 1525 by Giovanni da Verrazzano, who had poked his way all the way up the east coast of North America seeking an open waterway that would transport him directly into the Pacific Ocean. Hudson had already tried heading eastward from England over Russia but had given up and turned back. His trip to America was supposed to be another try via the eastern route, but, inspired by what Verrazzano had attempted, he fooled his backers and turned around to seek the opening in North America. Who could blame him? Eager Europeans had already drawn fanciful maps showing

*Hudson is often misidentified as Dutch, because the Dutch financed this particular expedition and referred to him as Hendryk.

only a narrow neck of land in North America between the two great oceans.

Hudson saw a broad waterway ahead of him as he sailed through the harbor. Although this waterway appeared to be heading due north rather than westward, Hudson decided to keep going, on the hunch that somewhere he would find a left turn to the west. And then, at long last, he would have found the link into the Pacific.

By September 19, he had traveled about 150 miles due north, to the area just above present-day Albany, without finding anything except a lot of Indians eager to trade with the white men. The only sign of a left turn was a skinny river that could hardly lead to the great Pacific Ocean. He ignored it and followed on up the northbound river. After Hudson and his crew had spent two nights amusing themselves by getting the local Indians drunk on wine and aqua vitae—which the Indians called "hoochenoo," thereby contributing the word "hooch" to the English language—Hudson sent a contingent of crew members ahead in a small boat to sound for depth.

They brought back bad news. The river became increasingly shallow as they rowed north. After about twenty-five miles, it was only seven feet deep. Hudson gave up, turned around, and headed the *Half Moon* back to Europe.

* * *

The very point where Hudson's crew found the river too shallow for further navigation would turn out to be the key, not to the nonexistent Northwest Passage, but to a very different western passage destined to play a far greater role in the world's history than the silks and spices of the Far East on which the European explorers like Hudson had set their hearts. De Witt Clinton would have known better—he was born and brought up in the Hudson River valley.

If Hudson had turned left instead of turning around, he would have seen what De Witt Clinton would be fighting for almost exactly two hundred years later, when the Erie Canal was being so fiercely debated. With a left turn, Hudson would have entered the mouth of the Mohawk River,* which runs due west for more than a hundred miles deep into New York State near the shores of Lake Ontario, the easternmost of the Great Lakes. He would have found the Mohawk valley to be the perfect portal to the vast lands and immense riches of America's west, unmatched by any other route along the stretch of more than a thousand miles of towering mountains that fill the space between the northern border of the United States and the southern end of the Appalachians near Atlanta, Georgia.

The Mohawk and the surrounding countryside played a critical role in the development of the Erie Canal. The distinguishing feature of the Mohawk valley is a deep gorge rising five hundred feet above the water and squeezing the river down to its narrowest point as it carves its way through a preglacial divide in the mountains. As a consequence, the Mohawk is the only river that slices right through the Appalachian mountain system. All the other rivers rise on the eastern slopes of the mountains and flow down toward the Atlantic, requiring land travel across rugged mountain passes to reach the western slopes.

Indeed, as an official geological history of New York State pointed out in 1924, "Had it not been for the recent [that is, the recent hundred million years or so by geological standards of measuring time] cutting of this gorge through the barrier at Little Falls, the Mohawk valley would never have been so important as a great gateway between the Atlantic coast and the west."[1] That is an immense understatement. Had it not been for the cutting of that gorge, the entire history of the United States would have been very different. Without that gorge, Americans might well have grown up split by the Appalachian mountain range into two quite separate nations, or perhaps into even more than two nations.

*The Mohawk River is named after the local Indians, but the word "Mohawk" means "cannibal" in their language.

That gorge made the Mohawk valley the nodal point of the national network of transportation, ideas, and economic and political forces forging the powerful link between the Atlantic states and the vast lands to the west. Even today, the main east-west railroad line, the New York State Thruway, and countless telephone wires and communication cables accompany the Mohawk River through this gorge.

* * *

The first man to recognize the almost unlimited potential of the route to the west provided by the Hudson and the Mohawk valleys was an Irish-born scientist, physician, and surveyor of Scots descent named Cadwallader Colden, who in 1724 was serving as surveyor general to William Burnet, captain-general and governor of the province of New York. Tall, with a round face and a protruding nose, Colden must have been a busy man indeed. Before coming to America, he had published works in natural history, applied mathematics, botany, medicine, and philosophy. His talents had led the great astronomer Edmund Halley to read his paper "Animal Secretions" before the Royal Society in London in 1715.[2] He was also an ill-tempered critic of the state. In 1748 he complained that "the only principle of Life propagated among the young People is to get Money."[3]

Colden was appointed lieutenant governor of New York in 1761 but soon fell out of favor with the citizens by refusing to support their popular agitation against the Stamp Act, the first provocative step taken by the English along the path that led to the American Revolution. Colden never changed his mind. Shortly before he died in 1776, he was still declaring his loyalty to King George III.

In 1724, Burnet sent Colden westward to report back not only on the nature of the territory in western New York and beyond, but also, and more important, on the activities of the French and the Indians, both political and military. The almost constant warfare between Britain and France would continue for another ninety years and spill over as a powerful force on developments in North America, militarily, economically, and politically.

Colden sensed the critical importance of this struggle. His account of his trip in 1724 is a remarkable document.[4] Titled "A Memorial Concerning the Furr-Trade of the Province of New-York" and nearly six thousand words in length (this chapter is approximately five thousand words), Colden's report to Governor Burnet provides a detailed description of the country he visited, including the topography and geography, the waterways, and the climate. He gives special attention to the navigable features (or lack of them) of the St. Lawrence River compared with the rivers and lakes between New York City and Lake Erie. In addition, he provides a meticulously detailed history of the province of New York and the unstable relationships between the Iroquois Indians and the other Indian tribes, as well as between the Indians and the French and British living and working in the province. As a bonus, he offers an insightful analysis of the economics of the fur traffic and the limitations of French trade policy, with recommendations for how the English could capitalize on the situation.

Colden's survey took him all the way to the Mississippi. When he stood on the banks of the great river, he was in awe of the scene before him. "There is opened to the view such a scene of inland navigation as cannot be paralleled in any other part of the world," he reported to Burnet.

Yet the English had no ready access to that unparalleled scene of inland navigation. The French and their Indian allies could boast that the bounds of New France reached from the mouth of the St. Lawrence River, which ran at a great northeastern arc from Lake Ontario to the Atlantic north of Maine, all the way to the mouth of the Mississippi on the Gulf of Mexico. The fabulous riches in furs and timber were controlled by a French monopoly. Most of the trade to France moved from the interior to the St. Lawrence.

Nevertheless, Colden was an optimist about New York. His report to Governor Burnet emphasizes over and over how generously the geography of British territory provides the key to an easy victory over the French. At heart a surveyor, he favors the gentle landscape and more navigable rivers in English territories in North America while the French "labour under difficulties that no art and industry can improve." The St. Lawrence is their only link to the Atlantic, over which they must carry the exports and imports for their trade with the Indians, but the river lies far to the north and is subject to "tempestuous weather and thick fogs . . . the navigation there is very dangerous. . . . A voyage to Canada is justly esteemed much more dangerous than to any other part of America."

Furthermore, the course of the river between Quebec and Montreal has tides of eighteen to twenty feet, currents often too strong for boats to make any headway, banks now wide and suddenly narrow, and so many sunken rocks "that the best pilots have been deceived." Travel at night is impossible. As a result, the French never attempt more than one voyage a year to Europe or the West Indies.

Now Colden clinches his argument. Thanks to the Hudson River, the English enjoy a transportation system far superior to the forbidding St. Lawrence. The Hudson is "very straight all the way [into northern New York State and the mouth of the Mohawk], and bold, and very free from sand banks, as well as rocks; so that the vessels always sail as well by night as by day, and have the advantage of the tide upwards as well as downwards." Furthermore, the distance from the great ocean port of New York City to Albany is only 150 miles.

That is just the beginning. As one later traveler put it, the mountains "seem to die away as they approach the Mohawk River."[5] Only about 110 miles of the rolling country of western New York State separate Albany on the Hudson from the headwaters of the Mohawk near Lake Ontario. At that point, the traveler has already covered 40 percent of the way to Lake Erie, with 230 miles, as the crow flies, of relatively flat country ahead. The elevation above sea level at Lake Erie is not much more than 100 feet above the level at the origin of the Mohawk.

The current on the Mohawk is much gentler than on the St.

Lawrence, with only two waterfalls of any size to interrupt the trip, although the Mohawk has occasional blocks from shifting sandbars or low waters. West of the Mohawk, Colden goes on to relate, passages by canoe are available all the way to Lake Erie along a straight route punctuated with Indian communities; these streams are separated from one another by only short distances, or "portages," over which canoes and cargo have to be carried. Lake Erie, in turn, opens the way through the connecting Great Lakes to the headwaters of the Mississippi as well as to contact with countless additional Indians living throughout the Great Lakes country, Indians who were eager to swap "peltry" and timber for ships' masts in exchange for woolen goods and the other enticing products white men had to offer.

Although not thinking explicitly in terms of an artificial waterway, Colden is certain of the case he has made: "It is only necessary for the traders of New-York to apply themselves heartily to this trade, for in every thing [such as] diligence, industry, and enduring fatigues, the English have much the advantage of the French." Dazzled not only by the political and military advantages of this territory but also its commercial potential, he asks, in pained astonishment, "How is it possible that the traders of New-York should neglect so considerable and beneficial trade for so long a time?"

* * *

Colden's British colleagues never provided a good answer to his question. They scattered a complex of military outposts throughout western New York and up into Canada, which would figure in the French and Indian Wars as well as in the American Revolution and the War of 1812.* But aside from a few intermittent efforts to improve the navigation of the Mohawk River, the English did nothing to capitalize on the

*One of these structures, of uncertain date, was named Fort Pentagon, anticipating by over two hundred years the name of a rather larger and more imposing North American military establishment.

enormous opportunities in developing the transportation system of the province of New York. The most significant of these proposals came from Sir Henry Moore, governor of New York in 1768. Moore had proposed to the legislature "the improvement of our inland navigation as a matter of the greatest importance to the province, and worthy of their serious consideration."[6] As would happen on future occasions, political infighting broke out on unrelated matters and Moore's recommendation came to naught.

But Colden had no doubt that the Mohawk was the key to the whole strategy. Much of the river is smooth sailing. Over its first forty miles from its origin, the river slopes downward only about a foot a mile. Then it encounters a waterfall with the unimaginative and common name of Little Falls, which drops more than forty feet in the vicinity of that crucial gorge. Below Little Falls the landscape tends to flatten out as the river covers another stretch of sixty-two miles with a decline of only two feet per mile. The greatest obstacle is close to the very end, where the river drops more than two hundred feet over sixteen miles between Schenectady and the point where it empties into the Hudson, just north of Albany. Over the entire length of the river, however, including the two big waterfalls, the downward slope amounts to only about four feet per mile.

Coming up from the Hudson, the first notable feature Colden would have encountered was the majestic cataract at Cohoes Falls (more grandly known among the locals as Great Falls), just a short distance upstream from Albany—and marred in the modern world by a line of water-powered factories and slumlike dwellings along the banks of the river. Here the water cascades downward seventy feet between steep rocks that continue to hem in the river for a great distance below the falls. An early Dutch visitor named Reverend Megapolensis, who visited the falls in 1642, described the waters as "boiling and dashing with such force . . . that it was all the time as if it were raining."[7] Two other Dutchmen, arriving on the scene about twenty years later, related how "you can hear the roaring which makes everything tremble, but on reaching [the falls] and looking at them, you see something wonderful, a great manifestation of God's power and sovereignty of His wisdom and glory."[8] No wonder the name of the falls is derived from the Algonquian word "*gahaoose*," which means "shipwrecked canoe."[9]

Once past the falls, Colden would have reached what was known in those days as S'coun-ho-ha—which the Dutch could say only as Schenectady, which stuck. The word meant the "door" to the Mohawk tribe that lived there, referring to the eastern entrance to the "long house," as they described their lands between the Hudson and the Finger Lakes south of Lake Ontario. In fact, when the French first visited the Iroquois Confederacy from the north in the 1680s, they were confused when the Iroquois asked them why they came "from the chimney instead of the door."[10] Around fifty miles farther west, just about midway on the Mohawk between its source and Albany, Colden would have reached Little Falls. Between Cohoes and Little Falls and then all the way up to the origin of the river near Lake Ontario, the land is green, the slopes are gentle, and the forests are lush.

When Colden visited the area, the countryside along the shores of the river was packed with forests where wild creatures such as eagles, wildcats, and beavers roamed freely. As late as 1793, a traveler noted, "The banks are lined with trees, which lean over like an arbor, so as to render the navigation shady and very agreeable, and it seems like being in a garden."[11] Forty years later, the famous English actress Fanny Kemble reported in the diary of her travels in America that "the valley of the Mohawk, through which we crept the whole sunshine day, is beautiful from beginning to end; fertile, soft, rich and occasionally approaching sublimity and grandeur in its rocks and hanging woods. We had a lovely day, and a soft blessed sunset . . . where the curved and wooded shores on either side recede, leaving a broad smooth basin [with] one of the most exquisite effects of light and colour I ever remember to have seen over the water and through the sky."[12]

* * *

In his report to Burnet, Colden's frequent reference to the Indians and their uncertain allegiances was a constant reminder that more than geography was at stake in the struggles with the French or the design of economic development in the area.

The history of the Mohawk valley is full of colorful Indian tales and myths, as well as bloody battles among the Indians, between Indians and whites, and among the whites themselves. When the first Europeans arrived in the early sixteenth century, the Indians in the valley were organized into the Iroquois Confederacy, a firm alliance of five nations—Mohawks, Oneidas, Onondagas, Cayugas, and Senecas—with headquarters near the headwaters of the Mohawk (a sixth nation, the Tuscarora, joined in the 1720s). The Iroquois controlled a wide area astride the lands between Albany and the Great Lakes to be crossed in the future by the Erie Canal.

Long after the Indians had been defeated and decimated by the whites—in 1800 a visitor to the area saw them as "poor enervate creatures [and] contemptible"—the great nineteenth-century American historian Francis Parkman would compare the Iroquois to the Romans because of their guiding principles of "peace, civil authority, righteousness, and the great law."[13] But in an important respect the Iroquois were superior to the Romans, because they formed a United Nations among themselves instead of one tribe conquering all the others.

For decades before the white men despoiled their civilization, the Iroquois sealed their unwritten agreement by casting their clubs and tomahawks into a pit, from which we derive the expression "burying the hatchet." The individual tribes were free to live by their own local customs, but their binding covenant provided for a lasting peace among the six tribes, an explicit procedure for settling disputes, and a united front in dealing with other Indian tribes. Their sophisticated political system included two legislative houses and a Grand Council—a structure admired by Jean-Jacques Rousseau in his concept of the "noble savage," as well as by George Washington and the framers of the American Constitution. In the heat of controversy at the Constitutional Convention, Benjamin Franklin saw fit to complain, "It would be strange if ignorant savages could execute a union that persisted for ages and appears indissoluble, yet like union is impractical for twelve [*sic*] colonies to whom it is more necessary and advantageous."[14]

When the Europeans arrived in the Mohawk valley, they were impressed with the political and military accomplishments of the Iroquois, but they were most interested in the shell beads the Indians were exchanging among themselves in elaborate ceremonies. The Indians called these beads "wampum." The wampum beads came in many shapes and sizes but with one notable feature in common: extraordinary hardness and durability. It was curious to find seashells in circulation so far into the interior, but over the centuries many items moved in the process of trade. Prehistoric Indians in Ohio had burial mounds containing items that could have come only from the Pacific Northwest.[15]

The Europeans immediately saw an opportunity to open a whole new world for wampum: money. The durability, uniform sizes and shapes, and broad acceptance of the shells made them an ideal vehicle as a means of exchange and store of value. The Indians' delight with gifts of wampum soon paved the way for more exciting purposes, such as swapping it for furs and other commodities. A few Indians were even yielding their lands to the Europeans in return for wampum.[16]

Metal coins were so scarce in the years leading up to the American Revolution that wampum rapidly evolved into a formal currency for transactions among Europeans as well as between Europeans and Indians. Official money is usually a monopoly of the state, but even in the forests of the Mohawk valley in the seventeenth century, supply could rise to meet demand. The Europeans became a kind of central bank, using their slender metal drill bits to produce vast quantities of wampum belt currency on a mass-production basis along the coast of southern New England. Before long, these beads were circulating as money from New York all the way westward to the Great Plains and southward to Virginia.

* * *

If the Mohawk were to be the perfect portal to the lands to the west, some kind of route had to be found between the marshy lands at its headwaters and Lake Erie about 230 miles away. This area was about as different from the dramatic gorge at Little Falls as can be imagined, but its curious geography would also prove to be just as important as the Little Falls gorge in the ultimate development of the area.

From the air, the headwaters of the Mohawk look like the random wanderings of a small child's crayon on a piece of paper. Countless small curves follow one another without any kind of pattern, some so compressed they look like horseshoes as they back around on themselves. In 1730, an unidentified group of people dug out a shortcut across the neck of one of these curves to simplify the navigation, a bit of engineering that can claim to be the first canal to be constructed in the New World; their handiwork is identified on later British maps as "the neck digged through in 1730."

There was a big payoff in reaching the land around the headwaters of the Mohawk, just west of the town of Rome. From there the route was open to Oneida Lake, a large body of water less than thirty miles from the "neck," easily crossed by boats and conveniently lying on an east-west axis. Before reaching the lake, however, the traveler had to contend with a meandering waterway known as Wood Creek, which rises on the other side of a small hill less than two miles away and flows in a westerly direction. The Indians called the short stretch of land separating the Mohawk River from Wood Creek De-o-wain-sta, or the Great Carrying Place; Europeans referred to it as the Oneida Carry or the Wood Creek Carry and built a short road between the two points in 1702.

Wood Creek at that point was little more than a twisting brook, so narrow in many places a man could easily jump across it. This was hardly the ideal waterway to transport people and cargo back and forth between east and west. A traveler passing through there in 1791 complained bitterly:

> The windings are so sudden and so short, that while the boat was ploughing in the bank on one side, her stern was rubbing hard against the opposite shore. . . . The boughs and limbs were so closely interwoven, and so low, as to arch the creek completely over, and oblige all hands to lie flat. These obstacles, together with sunken logs and trees, rendered our progress extremely difficult, often almost impracticable.[17]

Nevertheless, after about twenty miles of this aggravation, Wood Creek empties into Oneida Lake; the lake drains at its western end into a river that flows northward thirty miles into Lake Ontario at a place called Oswego.

* * *

Until the completion of the Erie Canal in 1825, the Mohawk–Wood Creek–Oneida–Oswego route was the only water route to the western parts of New York "known and open in practice," to adopt a phrase of Thomas Jefferson's. "Open" was something of an exaggeration. A voyage from Albany to Lake Ontario included major obstacles at Cohoes Falls and Little Falls, then the Wood Creek Carry, and, finally, swiftly running waters on the thirty miles from Oneida Lake to Lake Ontario. Altogether, the traveler heading west by bateau would need about two weeks to cover the two hundred miles from just above Cohoes Falls to Oswego on Lake Ontario.

Even after reaching Lake Ontario, the best route west to Lake Erie and the territories of Ohio and the Mississippi valley was still open to question. Westward connections from the lake were awkward, as the formidable Niagara Falls at its western side blocked direct access to Lake Erie and necessitated a difficult portage.

Despite all its problems, the Albany-Oswego route gained rapidly in importance. In 1727, only three years after Cadwallader Colden had returned from his trip to the Mississippi, the British took a small piece of his advice and established a trading house to do business with the Iroquois at Oswego. Two years later a large fort was built there to protect the fur traders, the first step in a trading perimeter the British could defend against the French. By 1730, the "Laws of the Colony of New York"—a continuing account of important events—could report, "Trade with the more remote Nations of Indians at the Trading House at Oswego is . . . Considerabelly [*sic*] Encreased."[18]

* * *

For the next fifty years, the Mohawk valley would be the stage for bloody sideshows of the continuing struggles in Europe between the British and the French. On occasion during their own conflicts, the French or the British (and their colonial associates like George Washington) would be at war with the Indians as well.

One day everything changed. On December 5, 1782, King George III went before the House of Lords to declare, "My lords and gentlemen: I lost no time in giving the necessary orders to prohibit the further prosecution of offensive war upon the continent of North America. . . . I did not hesitate to go to the full length of the powers vested in me, and offered to declare [the colonies] free and independent states." The American Revolution was officially over and a new nation proclaimed.

Almost two hundred years had passed since the first Europeans ventured into the Mohawk valley. Over all that time, the valley and the great space beyond it stretching to the Great Lakes remained a land of dark forests, roaring waterfalls, and meandering streams, of log cabins, little towns and military posts, of beaver pelts, wampum money, bitter battles at isolated forts, and, on occasion, white men's scalps hanging from Indian belts. With the single exception of the little neck cut into Wood Creek in 1730, the valley's natural features remained untouched. The ambitious men and women who settled in the valley were still colonials, denied a vision beyond their hunger for land. It was a place with no sense of the future.

Once there was liberty, the future was boundless. Cadwallader Colden's prediction of 1724 of "considerable and beneficial trade" would be just the beginning. A vision much greater than Colden's lay just ahead.

CHAPTER 3

WASHINGTON'S PIVOT

Despite Colden's enthusiasm for the Mohawk route to the west, the first person to take the risks and seek the rewards of linking east to west was not a New Yorker. A lot of talk would emanate from the New Yorkers in the early years after the victory at Yorktown, but the deeds came elsewhere.

While the New York legislature in Albany would engage in endless debates from 1784 to 1817, taking only tentative steps forward, and authorizing repeated exploratory commissions, a Virginian led the way by cutting a navigable waterway to the lands of the west in the early years after the Revolution. That this particular Virginian was very wealthy, had a keen eye for real estate, was a first-class surveyor, had proved his skills as a master administrator, and enjoyed unbeatable political connections certainly helped. Yet none of those advantages would have mattered without the single-minded conviction that his country's entire future rested on the successful completion of this project.

* * *

George Washington was first in the hearts of his countrymen, but he was also first to recognize the towering importance of a waterway to the west, first to arrange the financing, first to accomplish the necessary surveys, and first to supervise the complex engineering task it involved.

Washington never ceased to worry that the day might come—and sooner rather than later—when the ever larger numbers of people migrating into the vast territories west of the Appalachian Mountains would break away from the thirteen states along the Atlantic and form a separate nation. Most of these individuals had just picked up their belongings and moved out, leaving neither family nor economic bonds behind them.

He had begun an effort as early as 1775, interrupted by the Revolution, to convert the Potomac River into the primary link to the west. In 1785, after completing a trip all the way from Mount Vernon to Pittsburgh and then down the Ohio River, he expressed his urgent concerns to the governor of Virginia:

> I need not remark to you, Sir, that the flanks and rear of the United States are possessed by other powers—and formidable ones too: nor need I press the necessity of applying the cement of interest to bind all parts of the Union together by indissoluble bonds—especially of binding that part of it which lies immediately west of us, to the middle States. For what ties let me ask, should we have upon those people, how entirely unconnected with them shall we be, and what troubles may we not apprehend, if the Spaniards . . . and Great Britain . . . should hold out lures for their trade and alliance? What will be the consequence of their having formed close commercial connexions with both or either of those powers? . . . The western settlers (I speak now from my own observations,) stand, as it were, upon a pivot. The touch of a feather would turn them any way.[1]

But even worse than breaking away, the westerners might choose to join up—or be forced to join up—with the Spaniards, the French, or the English Canadians, who still controlled most of North America and who together surrounded the newcomers on the south and the north while intermingling with them in the west. However the rupture occurred, it would leave the United States just a small country clinging to the ocean shores and hemmed in by a much larger western neighbor of dubious reliability.

The manner in which Washington expressed his concerns foreshadows in a remarkable way how De Witt Clinton would argue the case many years later in his long effort to transform the Erie Canal from a dream into a reality. We do not know whether Clinton ever met George Washington in person. He was only fourteen years old when Washington was cheered in the victory parade in New York in 1783, but Clinton's father, James, a professional soldier and a general in the Revolution with a distinguished record in battle, was on horseback just behind Washington in the independence parade.

In 1816, in a celebrated document that would finally persuade the New York legislature to proceed with building the Erie Canal, Clinton wrote, "However serious the fears which have been entertained of a dismemberment of the Union by collisions between the north and the south . . . the most imminent danger lies in another direction. [A] line of separation may be eventually drawn between the Atlantic and the western states, unless they are cemented by a common, an ever acting and a powerful interest." He emphasized how "one channel, supplying the wants [and] increasing the wealth . . . of each great section of the empire, will form an imperishable cement of connection, and an indissoluble bond of union."[2]

* * *

Washington receives too little credit, among his many accomplishments, for having a keen head for business. He was in full agreement with how the French foreign minister, Charles Gravier, had characterized the Americans in 1785: "These people," he observed, "have a terrible mania for commerce."[3] In Washington's search for a way to forestall the ominous possibility of a breakup of the union as a result of the mountain barrier, he had come up with a remarkably modern solution that anticipated Clinton's view of the matter over thirty years later. He was convinced that trade and commerce—and only trade and commerce—could bind the western territories to the United States forever with a viable sense of community and mutual interdependence. And there was not a moment to be lost.

Economic linkages, however, depend on geographical linkages. In Britain, France, and the Netherlands, the various regions were joined to one another with networks of waterways—rivers, canals, or both—capable of moving large quantities of cargo all around the country and to and from the seaports. The United States would never develop into a great nation as long as most of its citizens clung to the seacoast while the more adventuresome drifted off to the west, separating themselves from the seaboard by the formidable chain of mountains running down the entire country. It was absolutely essential to replace the rutty dirt roads and Indian paths now in use with a system of internal waterways through the mountains and into the west. But where and how?

Washington's answer to this question had complete support from his friend Thomas Jefferson. Although they approached the matter as prominent Virginians as well as patriots of the new United States, it could hardly have been a coincidence that Washington also happened to be the largest landowner on the Potomac River.

"Nature then has declared in favor of the Potomac," Jefferson wrote to Washington in early 1784, "and through that channel offers to pour into our lap the whole commerce of the Western world. But unfortunately [the route] by the Hudson is already open and known in practice; ours is still to be opened."[4] To which Washington responded, "With you I am satisfied that not a moment ought to be lost in recommencing this business [of opening the Potomac], as I know the Yorkers will lose no time to remove every obstacle in the way of the other communication."[5]

Now a squire liberated from military and, for the moment, political obligations, Washington could turn his attention to the future of his own fortune as well as the future of his country. As the exchange with Jefferson suggested, he recognized the attractions of building a water route through the Mohawk valley and beyond—and the huge impact such a development would have on property values. Only a few months earlier, he had traveled across New York State as far west as Lake Erie in the company of the state's first governor, George Clinton, perhaps to seek new lands to add to his already extensive properties in Virginia and Pennsylvania.

Washington was deeply impressed with the potential of the area. "Prompted by these observations," he wrote, "I could not help taking a more contemplative and extensive view of the vast inland navigation of the United States, and could not but be struck by the immense diffusion and importance of it . . . and with the goodness of that Providence which has dealt his favors to us with so profuse a hand. Would to God we may have wisdom to improve them. I shall not rest contented until I have explored the western country, and traversed those lines . . . which have given bounds to a new empire."[6] At about the same time, with remarkable foresight, Washington referred to New York as "the seat of empire."[7]

Binding the westerners to the Atlantic communities was a goal that would consume Washington for the rest of his life. As late as 1796, in the course of his Farewell Address, he was still warning against the dangers facing the separate parts of his country unless they could forge an indissoluble community of interest as one nation: "Any other tenure . . . whether derived from its own separate strength, or from an apostate and unnatural connexion with any foreign power, must be intrinsically precarious."

* * *

In September 1784, six months after the correspondence with Jefferson on these matters, Washington set out on horseback to look over his vast landholdings for the first time since the outbreak of the Revolution in 1775. As he wrote in his diary before leaving, a not incidental objective of the trip was to plan, "as much as in me lay, the inland navigation of the Potomac."[8]

At a time when the accumulation of financial assets as a form of wealth was virtually unknown, Washington had a passion for real estate. During the years leading up to the American Revolution, starting as early as the age of eighteen, he had accumulated 5000 acres in Virginia, comprising eleven large properties scattered all the way from his family home at Mount Vernon to the northwestern reaches of the Potomac River valley. Mount Vernon itself comprised another 8000 acres. He also owned six even larger properties in the Ohio River valley in Pennsylvania, adding up to 33,000 acres, which he had received in 1770 as a reward from a grateful British government for his outstanding service as a lieutenant colonel in the French and Indian Wars. When the British let Washington select the lands to be given to his men as well as his own, he described his grants as "the cream of the Country in which they are; that they were the first choice of it; and that the whole is on the margin of the Rivers and bounded thereby for 58 miles."[9] He advertised these properties for lease after the Revolution, promising that "a great deal of it may be converted into the finest mowing ground imaginable, with little or no labour."[10]

The route Washington was to follow in 1784 was a familiar one, not only to him but to many others heading to the west. As early as 1745 an enterprise called the Ohio Company had carved out a trail by following along the natural river as far as Alexandria, close to what would one day be Washington, D.C., and then clearing wagon trails around the two large sets of falls just above Alexandria. Beyond that point, they returned to the river for the 160-mile trip right up to the foothills of the Allegheny Mountains at Fort Cumberland, now Cumberland, Maryland, where the Braddock Road led overland to the Monongahela.[11]

After Washington had ridden for about 125 miles from Mount Vernon, he stopped at a little town named Bath, set in the midst of the therapeutic warm springs of the area. This part of the country had been familiar territory to Washington for many years. In 1748, when he was only sixteen years old but already a skilled surveyor, he had been invited to participate in a survey of the five million acres of property throughout Virginia held by Thomas, Lord Fairfax, the only British peer whose primary residence was in America. These far-ranging properties, originally granted by King Charles II, would be liquidated by the family after the Revolution, between 1781 and 1808.

Lord Fairfax himself had founded the town of Bath in 1776, when Virginia was still an English colony, as a small but ambitious replica of Britain's famous spa of the same name. Almost immediately afterward, Washington joined with friends from the colonial elite to buy properties in Bath and participate in creating the country's first fashionable spa, "Ye Famed Warm Springs," as he had described the area while conducting his 1748 survey.[12]

* * *

Now, in 1784, Washington stopped at the Inn at the Liberty Pole and Flag, a new establishment advertised as "a very commodious boarding house for the residence of ladies and gentlemen."[13] The innkeeper, James Rumsey, was a large and handsome man with a sensitive, almost feminine mouth, and fully conscious of his own charms. He lived high, relished risk taking, and was accustomed to carry on in the face of failure with no evident loss of momentum or enthusiasm.

Washington fell under Rumsey's spell and in short order gave him a contract to build two new houses on his properties in the town. But Rumsey had more than real estate deals in mind to discuss with his distinguished guest. Describing himself without hesitation as a genius of an inventor, Rumsey confided to Washington that he had concocted a mechanical boat capable of traveling upstream against the natural flow of a river. Washington was incredulous, since no one had ever succeeded in creating such a device.

Rumsey had not constructed a full-size working version of his invention, but he did have a miniature model set in a large shallow pan through which water could flow rapidly. The mechanical boat was a kind of catamaran, consisting of two separate vessels joined with a paddle wheel between them and a long pole reaching from each of the outer sides of the contrivance down to the bottom of the pan. The water flowing through the pan turned the paddle wheel, which in turn moved the poles in such a way that they propelled the entire contraption forward. In this fashion, the little model walked itself forward against the current of the water.

Rumsey's mechanical boat looked like a gift from heaven. He had dared to defy one of nature's most daunting features. By propelling a boat upstream, Rumsey's apparatus could change the world. Washington immediately grasped its critical role in binding east and west by creating a two-way route on the Potomac River, a route that not incidentally would run right through his own lands before crossing over to the vast spaces on the other side of the mountains.

All the rivers in Virginia flow eastward from the mountains down to tidewater. That was fine for transporting the output of agriculture, furs, and timber from the western lands downstream to the Atlantic states. But unless they could find a means to move boats against the current up to the formidable Appalachian chain, the United States could never supply the westerners with everything they would require from the factories along the ocean shores or from the imports arriving at the Atlantic ports from abroad. The Indians' elegant, lightweight bark canoes were adequate for carrying small cargoes upstream, but they would never do for the potential volume of traffic that could develop with more modern means of transportation. Without a two-way connection by water between the Atlantic seaboard and the other side of the mountains, the people in the west would have no choice but to use the rivers running through Spanish, French, and Canadian territory. Then the whole future of the United States would be at risk.

Washington immediately prepared a certificate for Rumsey describing this marvelous and unique invention as "of vast importance, maybe of the greatest usefulness" for the purpose of moving a boat upstream.* At the same time, he noted in his diary that Rumsey's mechanical boat would "render the present epoch favorable above all others for securing . . . a large proportion of the product of the western settlements, and of the fur and peltry of the Lakes also . . . binding these people to us by a chain that can never be broken."[14]

Rumsey would leverage his invention, and Washington's sponsorship, into an extended effort over the years to invent the steamboat, but that is another story.†

* * *

Washington's only obstacle to establishing the perfect link between the Atlantic states and the western lands was to transform the Potomac from a series of rocky rapids and frequent waterfalls into a navigable waterway. Although he was well aware of how much had to be done before the dream would be a reality, and although he was enjoying the freedom of retirement, he began to give the matter serious consideration because of its overwhelming importance. Through all the mishaps that lay ahead, he remained convinced of the superiority of the Potomac to "open and make easy" the future gateway to the west. "The navigation of this river is equal, if not superior, to any in the Union," he boasted. "This will become the great avenue into the Western Country."[15]

Washington was careful to avoid any hints of self-interest in his zeal to promote the development of the Potomac. On the occasion of a social visit from his old comrade-at-arms the Marquis de Lafayette, he gave a speech in which he insisted he had no motive for promoting the devel-

*Here, too, Washington and Jefferson were in enthusiastic agreement: Jefferson would later characterize Rumsey as "the most original and the greatest mechanical genius I have ever seen."

†And a great story it is! Washington was to figure in this story as well. For the full history, I heartily recommend James Flexner's *Steamboats Come True*, a great read on many levels.

opment of the Virginia waters other than the advantage it would provide to the entire nation—and Virginia in particular. The project, he declared, would cement east and west while providing important stimulus to passenger travel and commercial activities. In his letter to the governor of Virginia after his trip west in 1784, Washington suggested the immediate appointment of commissioners "of integrity and abilities, exempt from the suspicion of prejudice" to seek the best kinds of connections, by water or by land, to the Ohio River, which was the critical linkage to the whole transportation system he had in mind.

But Washington would soon be putting his money where his mouth was.

* * *

Here again, Jefferson cheered Washington on. In a remarkably contemporary vein, Jefferson wrote to Washington, "A most powerful objection always arises to propositions of this kind. It is, that public undertakings are carelessly managed, and much money spent to little purpose. . . . [Would your] superintendence of this work break in too much on the sweets of retirement and repose? If they would, I stop here. . . . But what a monument of retirement it would be!"[16]

Washington went to work. The vehicle he selected for the business side of his enterprise had been organized twenty-five years earlier by a group of eager Virginians who shared his dream of making the Potomac navigable. They called their enterprise the Patowmack Company. Not much came of it. But now Washington took over the management of the Patowmack Company, convinced that the Potomac was the superior route to link the Atlantic states to the west and enthusiastic about the feasibility of using Rumsey's discovery to move boats upstream against the currents. He changed the company's name to the Patowmack Canal Company, had himself appointed president, and lobbied the states of Maryland and Virginia for a new charter. Finding the money to do the job and hiring a skilled engineer to carry it out was all that remained.

Washington was a natural as a money raiser. Requests from him were hard for anyone to turn down, just because of who he was, but also because of his great leadership abilities and his proven skills as a surveyor. If anyone could accomplish this task, and make it work, Washington could. After purchasing 73 shares for himself out of the 250 authorized shares in the new company, he rapidly sold the remainder to a willing group of venture capitalists.

The subscription price for the 250 shares was £50 each, the rough equivalent of $250—if there had been U.S. dollars in existence in 1784—for an anticipated total of £12,500 or $62,500.[17]* It may seem odd that Americans were still denominating transactions in pounds now that they were independent of Great Britain, but the establishment of a common American currency would not take place until after the establishment of the United States later in the decade and Secretary of the Treasury Alexander Hamilton carried out the transformation.

The amount involved here sounds piddling. In 1784, however, the American gross national product was only a tiny fraction of today's $10 trillion, and the population was less than 4 million people compared with 295 million today. And Washington's estimate was probably too low.

Selling these shares was not the same thing as raising the money. In those days, investors *subscribed* to shares, which meant they could pay on an installment basis rather than putting up all the money at one time. Most of the money Washington expected never arrived, forcing him to appeal repeatedly to the legislatures of Maryland and Virginia for help. The magic of his name made all the difference here too, and in this case the money came through: each state subscribed to—and paid for—fifty shares, or one-fifth of the total. Before they were done, the two states would contribute as much as $150,000, or just about half of the total cost of the project.[18]

Nevertheless, the negotiations with the state authorities were com-

*In 2005 dollars, this is approximately $1,125,000.

plicated, as for most of its length the Potomac forms the border between the two states. To iron out the uneasy relationship between the two entities, Washington invited them to attend a conference in March 1785 at Mount Vernon. The resulting Mount Vernon Compact, soon ratified by the Maryland and Virginia legislatures, was actually an illegal act, because the Articles of Confederation, under which the United States was then operating, prohibited treaties between individual states without approval of Congress. Although the participants recognized that the compact was technically illegal, it revealed the complexities of interstate relations under the Articles of Confederation and was the first defining event leading to the Constitutional Convention two years later.[19]

* * *

Even with these difficulties, raising the money was the easy part of transforming the Patowmack Canal into the smooth connection between the Atlantic states and the west. As a skilled engineer and surveyor familiar with the territory, Washington had every reason to believe he knew what he was doing. Yet the decision was an audacious one. It was not at all obvious that the Potomac River would be as submissive to his commands as he anticipated. Much as Washington loved the Potomac, the river became an implacable enemy, fighting against his plans for it almost every inch of the way.

The river wanders from the high point of the Blue Ridge Mountains down to the sea, alternating between stony, shallow waters and precipitous rapids or falls. Huge rocks emerge from the swiftly running currents, with soft riverbeds rare. The headlong rush of current along many stretches of the river insisted on shoving new rocks into the stretches that backbreaking dredging efforts had already cleared. About thirty miles up from the beginning of the canal at Alexandria, seasonal floods could push as much as a million gallons a second over Great Falls, leaving the work crews terrified and only too willing to abandon the massive construction job before them.[20]

Water travel for anything more than a canoe becomes impossible upstream to the west of Fort Cumberland. At this point, travelers and cargo would have to transfer to wagons and proceed by road, traveling overland on the rough 1775 Braddock Road in Pennsylvania and then heading downstream on the Monongahela River. The Monongahela, in turn, presented its own problems: in some stretches, it was little more than a creek, but after heavy rains it could rise as much as forty feet.

Washington had no doubts about his ability to overcome all of these obstacles. Before he could even begin to face up to this challenge, however, he had to find someone to take care of the day-to-day management of the job and help in the design and engineering plans as well. When he advertised for a man with sufficient knowledge and training to fill this role, no one showed up. Other small canals had already been constructed in the United States, but there was not a single individual in the entire United States at that moment with adequate hands-on experience to deal with a project as extensive and complicated as this one.

So Washington turned to the best man he knew under the circumstances: who else but the inventive innkeeper James Rumsey. Washington could see no problem in Rumsey's lack of knowledge about building canals or improving river navigation. "I have imbibed a very favorable opinion of your mechanical ability," he wrote to Rumsey, "and have no reason to distrust your fitness in other respects."[21] Rumsey in return had no hesitation in accepting the job and signed up on July 14, 1785.

Problems popped up from the start. The water in the Potomac was freezing cold as it poured down from the mountains and through the wilderness, forcing the laborers to toil in icy streams up to their thighs for hours on end. Rumsey's name was appropriate for the problem, because rum was all too often a necessary lubricant to keep the work in motion. Rum created its own problems, as Rumsey would write: "Every time they get a little Drunk, I am cursed and abused about their money in such a manner that contrary to my wish, I am obliged to turn abuser."[22]

So, for the most part, the Patowmack Company ended up employing workers unable to find gainful employment elsewhere; many of them were prisoners, rough customers who enjoyed creating mayhem in the surrounding countryside. With responsibility for about a thousand men, Rumsey grew desperate. "The complaints that was [*sic*] made to me was shocking," he wrote Washington, "that no person could come on their lawful business but what got abused, and the officers of justice durst not go on the grounds to execute their offices."[23] This situation left Washington with no choice but to buy black slaves and to take over the indentures of Irish immigrants who had mortgaged years of their future labor as a means of getting themselves to America, and add these men to the canal-building force. Many of them tried to run away from the horrors of the work.[24]

Obsessed with the noble goal of employing steam to drive a boat upstream, Rumsey was soon fed up with the constant aggravation of his responsibilities on the Patowmack Canal. He was also continuing his efforts to convert his model of a pole boat into a reality for moving cargo westward on the Potomac. His duty on the river itself increasingly took second place and sometimes even third.

The directors fired him in July 1786. Although Rumsey protested loudly and tried to blame the directors for the lack of progress, Washington himself described Rumsey's accusations as "malignant, envious, and trifling."[25]

* * *

Despite all these difficulties, as well as Washington's removal from active management in 1788 because of his promotion from president of the Patowmack Canal Company to president of the United States, the canal did become a functioning waterway. But nothing came easily, with either the men at work on the project or the daunting physical problems they had to resolve. The locks at Little Falls, Virginia, were completed in 1795, nine years after Rumsey's departure but two years before Washington would end his second term as president of the United States. The five locks at Great Falls were an even more herculean task that dragged on for another seven years. By the time the job was done, Washington had been dead for three years.

In the end, the Potomac won out over Washington. While corn, whiskey, furs, and timber floated downstream toward the Atlantic without much difficulty, only trifling amounts of light manufactured goods moved upstream toward the mountains. Rumsey's gadget for propelling boats upstream had long since been forgotten, and the early steamboats had no way of dealing with the Potomac's long and frequent passages of turbulent and shallow waters. Instead, crews literally pushed their boats up the river by sticking their poles in the bed of the river and then walking the planks on the sides of the boats as if they were on a treadmill.[26] At the most difficult places, even that effort failed and the boatmen had to pull their boats forward by holding onto ropes or chains fixed to the rocky sides of the river.

Most of the boats were little more than thin wooden rafts or crude affairs with flat bottoms and a steering oar, carrying about thirty or forty tons each. The crude character of the riverboats was neither an accident nor the result of a lack of boatwrights' skills. In view of the difficulty of making the return trip upstream, nearly all the vessels were built to be broken up at the end of their eastbound journeys for sale as timber, fuel, or wood for housing construction. Then, with their boats dismantled, the boatmen walked back home up the river valley.

The vision of a thriving two-way artery, firmly clasping east and west to each other, remained a dream.

Washington had never lost faith. On December 10, 1799, too ill to attend a company meeting, he voted his seventy-three shares in the company by proxy. Four days later he was dead.

His Patowmack Company died too, in bankruptcy, in 1810 after twenty-six years of operation, with all its assets and liabilities turned over to the new Chesapeake & Ohio Canal Company. But the C & O

Canal was destined to become nothing more than a tourist attraction; in 1828, the earth would be broken for a brand-new technological marvel, the Baltimore & Ohio Railroad.

* * *

In 1784, Washington had predicted to Jefferson that "the Yorkers will lose no time to remove every obstacle in the way of the other communication." He need not have worried. The New Yorkers appear to have been strangely unhurried about launching the great enterprise that would finally succeed in landing the nation safely on the happy side of Washington's pivot. By the time the Yorkers finally started work on their canal, George Washington had been in his grave for eighteen years.

As far back as 1724, Cadwallader Colden asked, "How is it possible that the traders of New-York should neglect so considerable and beneficial trade for so long a time?" The answer to his question comes to us right off today's headlines. Where private enterprise, entrepreneurship, and self-interest are in charge, events can move rapidly. Where politicians are running the show, decision making is diffused, struggles for dominance are endemic, and risk is something to be avoided rather than welcomed.

The story of the Erie Canal, and the stubborn and self-serving political wars that marred it, makes a stark contrast to Washington's Patowmack Company, where entrepreneurship and self-interest guided lofty and patriotic objectives. But it was the Erie Canal rather than the Patowmack Company that survived to dominate history.

CHAPTER 4

CANAL MANIACS

The Erie Canal was destined to have many fathers. Cadwallader Colden was perhaps the first to have the vision, but many others would have to follow in his footsteps before a project so immense and so novel for a country so young could reach a point where all the pieces would fall into place. Some who would appear at first to be solid champions of the canal would back away when the chips were down, while others who presented themselves as opponents would turn into avid supporters when most of the controversies had passed.

The most steadfast and effective of the canal's band of fathers turned out to be a motley group of men, ranging from the highest ranks of American society and political power to humble surveyors, merchants, and legislators. Their determination, and the means they employed to make the canal a reality, provide an important window into the early development of the American political structure. More important, these men have left us with vivid examples of the limitless enthusiasm for a bold future displayed by so many Americans in the early years of the nation—and even before it was a nation. And their achievement, when all of it was a reality, would exceed even their wildest and fondest dreams.

* * *

Nobody took any concrete action in response to Colden's "Memorial Concerning the Furr-Trade" for another sixty years. But matters began to progress more rapidly in the mid-1780s, at the same moment Washington was having fantasies about Rumsey and his mechanical boat. Three men in particular were pushing hard to make things happen. One got nowhere, although he had everything going for him. The second was more interested in words than deeds, but the words carried weight. The third was by far the most effective: he got the idea, looked the situation over, and pulled all the right strings to mobilize the kind of support that mattered.

The man who got nowhere even with everything going for him was Christopher Colles, a forty-six-year-old Irish mathematician, mechanic, and artillery officer in the Revolution. With identical alliteration in his name, Colles was, like Cadwallader Colden, a man of many talents. De Witt Clinton characterized Colles as "a man of good character, an ingenious mechanician, and well skilled in mathematics."[1]

Before emigrating from Ireland, Colles had been in charge of the engineering work on improving navigation on the River Shannon, which runs over two hundred miles from northern Ireland down to its estuary in the southwest—at a point where many travelers in the early days of commercial air travel across the Atlantic would enjoy duty-free shopping while their aircraft was being refueled. Colles was also familiar with the technological innovations of James Brindley and the Duke of Bridgewater.

He immigrated to America because he was convinced the talents and experience he had gained on the Irish and English waterways would be in high demand in the colonies. He first comes into view in 1772, when he delivered a series of public lectures in Philadelphia on the control of gases, including the advantages of an air pump of his own invention. He was in New York City the following year to give a presentation on the mechanics of canal navigation, based upon his experience on the Shannon and his studies of canal operations in Britain, including the Duke of Bridgewater's canal in particular.

In 1776, Colles designed a system of pipes made of hollow logs to transport water for New York City's water system from the major reservoir of the time (now under the Criminal Courts Building) to a subsidiary reservoir at the intersection of Broadway and White Street in lower Manhattan. This significant innovation was replaced about twenty-five years later when Aaron Burr obtained a charter for the Manhattan Company to provide twenty miles of wooden pipes supplying 700,000 gallons a day of water to 1400 New York homes.[2]* Burr's water company was immortalized by its transformation shortly afterward into the Bank of the Manhattan Company, set up at 40 Wall Street, an attempt by Burr to break Alexander Hamilton's Bank of New York and its monopoly of the banking business in the city.† As we shall see, Burr was also able to use the bank to further his own political ambitions.

Colles's achievement enjoyed a more gracious conclusion. On January 6, 1776, the City of New York financed his water system with an issue of £2750 of promissory notes, with the back of each note carrying an attractive engraving of Colles's proposal for a steam-powered water pump to drive the water to its final destination. As the notes appeared in denominations of two shillings, four shillings, and eight shillings, a large number of these pieces of paper must have been in circulation, advertising Colles's mechanical ingenuity.

No information is available on what originally provoked Colles's fascination with the Mohawk River and the Great Lakes, nor is there any indication that he was familiar with Cadwallader Colden's intense interest in the area. Nevertheless, Colles made a full-dress presentation to a committee of the New York State legislature in November 1784, arguing

*Daily consumption of water in New York City today is approximately 1.4 billion gallons.

†Burr's bank was called the Bank of *the* Manhattan Company because the Manhattan Company had been Burr's water enterprise, and the bank was an outgrowth of that.

for the rewards to be gained by improving the navigation on the Mohawk, "so that boats of burthen may pass the same."[3] He went on to point out that the ground between the upper stretches of the Mohawk and Wood Creek is perfectly level, providing an ideal channel to enter into the extensive country to the west.

Colles must have been convincing, although not as persuasive as he had hoped. After hearing the case, the chairman of the committee reported that "the laudable proposals of Mr. Colles . . . merit the encouragement of the public; but that it would be inexpedient for the legislature to cause that business to be undertaken at the public expense." Instead, the committee suggested, Colles should undertake the Mohawk project himself "with a number of adventurers (as by him proposed)," in return for which the State would grant him and his partners full rights in all profits from tolls to be collected on the river. Colles returned to the legislature six months later, in May 1785, and this time he scored a more impressive success. The legislature ordered Colles to prepare a full-fledged plan for presentation at the next session and, meanwhile, provided him with a payment of $125 (equal to about $2250 today) to evaluate the effort required to clear the obstructions along the Mohawk all the way up to its headwaters near Lake Ontario. Colles must have been compelling in his arguments.

He immediately set off to analyze the navigation of the Mohawk. The pamphlet Colles published on his return sets forth his case at length (no one in those days appears to have any sense of parsimony in the use of language), under the majestic title of "Proposals for the Speedy Settlement of the Waste and Unappropriated Lands of the Western Frontiers of the State of New York, and for the Improvement of the Inland Navigation Between Albany and Oswego."

Colles should surely have listed rhetoric among his many gifts. After a detailed description of what he describes as his "design," he goes on to assert, "By this, the internal trade will be increased—by this also the foreign trade will be increased—by this, the country will be settled—by this, the frontiers will be secured." After a further series of "by this'es," Colles completes his sequence with, "By this . . . if we please, the luxuries of life may be distributed to the remotest parts of the great lakes which so beautifully diversify the face of this extensive continent." The Great Lakes, he points out, have five times as much coast as all England, and the area watered by the rivers flowing into these lakes is "full seven or eight times as great as that valuable island." Easy access to the lakes was surely worth great effort.

The next session of the legislature, in 1786, passed a bill providing further compensation to Colles and his associates. This legislation, which focused on improving the navigation of the waterways right up to Oswego on Lake Ontario, included the prophetic phrase, "and for extending the same, if practicable, to Lake Erie."[4] But this part of the story has an unhappy ending. Colles failed to gather enough subscribers to launch a viable company capable of making the full length of the Mohawk River navigable, much less anything beyond the Mohawk. His vision crushed, Colles was left with no choice but to abandon his project.

He was not finished as an innovator, however. In 1789, he began work on the first road map or guidebook to the United States, consisting of an immense number of finely detailed strip maps, each of which covered an area of only about twelve miles. Three years later, Colles's extraordinary atlas had progressed to a point where it covered a distance of some one thousand miles from Albany, New York, to Williamsburg, Virginia. This book, considered a classic today, is the primary source for information on the road patterns and the development of transportation in the young republic. Yet, like his other great ideas, Colles's book was too far ahead of its time to catch on. By 1792, he had so few subscribers that he abandoned his atlas and faded into oblivion.

Colles's failure to generate more support for his Mohawk project is difficult to explain in view of his talents and achievements. But at a time when the supply of "adventurers" willing to provide risk capital was meager indeed, even George Washington had had difficulty in persuading his fellow investors to come up with the money they had promised him. It would take louder and more influential voices than Colles's to launch a scheme of this magnitude and complexity.

* * *

One of those influential voices was Gouverneur Morris, a powerful and prominent politician of great wealth, a passionate patriot, and a man given to grand eloquence in copious self-expression. Morris first spoke up in 1777, in the darkest days of the American Revolution, and even before Christopher Colles was pressing for improved navigation on the Mohawk. At that moment, General Philip Schuyler and his troops had been forced to abandon the strategic site of Fort Ticonderoga between Lakes George and Champlain and were in retreat to Fort Edward, about fifty miles to the south. General Washington responded by dispatching a Committee of General Safety, with Morris as one of the members, to meet with General Schuyler and report back on the condition of the troops and the overall strategic situation. Morris would turn out on occasion to be more interested in words than deeds, but his words would carry weight. This episode was only the first of his many associations with the conception and construction of the waterway across New York.

A vivid account of being in Morris's electrifying company during the inspection of the Committee of General Safety comes from Morgan Lewis, one of the officers in General Schuyler's demoralized headquarters in 1777. According to Lewis, Morris was a man "whose temperament admitted of no alliance with despondency, even in the most gloomy periods of the war, with which our situation might justly be classed; never doubting the ultimate triumph of our arms and the consequent attainment of our independence. . . . [He] frequently amused us by descanting with great energy on what he termed 'the rising glories of the western world.' "[5]

Lewis reports on one evening in particular, while Morris was "describing in the most glowing terms, the rapid march of the useful arts through our country, when once freed from a foreign yoke, the spirit with which agriculture and commerce, both internal and external, would advance; the facilities which would be afforded them by the numerous watercourses intersecting our country, and the ease with which they might be made to communicate; he announced in language highly poetic, and to which I cannot do justice, that at no very distant day, the waters of the great inland seas, would, by the aid of man, break through their barriers and mingle with those of the Hudson."[6] One wonders if Morris paused for breath in the midst of all that.

Just three years later, Morris would sum things up in his letter to a friend, "Shall I lead your astonishment to the verge of incredulity? I will: know then, that one-tenth of the expense born by Britain in the last campaign, would enable ships to sail from London through Hudson's river into Lake Erie."

Morris's enthusiasm for this project and his charismatic words never faltered over the years ahead, no matter how many obstacles others raised. Without his foresight, influence, and unqualified support, the canal might have taken even longer to become a reality. In time, he would cease to be just a rooter for the canal and would assume an active role in its development—but that moment was still ten years in the future.

* * *

Gouverneur Morris had been eloquent in expressing his dreams about a waterway across New York State. Both Colden and Colles had realistic visions of what had to be done. But it took the unremitting efforts of a man with the strange name of Elkanah Watson to produce the first tangible efforts to improve navigation in New York State since "the neck digged through" on Wood Creek in 1730.

Watson was in many ways as odd as his name. Born in 1758, he was many steps down the social scale from Gouverneur Morris, having started his career as an indentured servant for a wealthy local family. That background in no way inhibited him from making it his life's work to associate on intimate terms with the rich and famous. Indeed, Watson would surely rank high among the social climbers of history. He got off to a running start through his work as a courier at the highest levels of command during the Revolution, as well as his several trips to Paris carrying messages between Benjamin Franklin and the Continental Congress in Philadelphia. He numbered "Dear Dr. Franklin" among his closest friends, boasted a speaking acquaintance with the Marquis de Lafayette, entertained Hamilton and Burr (together) at his table, had frequent friendly encounters with James Watt, breakfasted "in a familiar manner" with Edmund Burke, and was introduced to the Prince of Wales (the future George IV) at the opera.[7] When King George III declared the colonies "free and independent states" in 1782, Watson was sitting in the House of Lords for the occasion, in the company of the famous American painters John Singleton Copley (who had painted a "splendid portrait" of him) and Benjamin West, as well as Admiral Richard Howe, the commander of the British Navy during the American Revolution.

Watson maintained a steady flow of correspondence with John Adams, who sent him a remarkable letter in 1823, addressing Watson as "My dear friend." Adams begins by summing up Watson's lifetime accomplishments, recalling "your meritorious exertions for the public good. Your active life has been employed, as far as I have known the history of it, in promoting useful knowledge and useful arts; and for which I hope you have received, or will receive, a due reward." But then, recognizing from his own experience how thin-skinned Watson was, Adams goes on to reassure Watson that "shafts are wanton sports, and secret and public malice are common to you, and all men who distinguish themselves. 'Envy does merit, as its shade pursue, And like the shadow proves its substance true.' This, or something more sublime, must be the consolation of all of us."[8]

Watson was also an insatiable tourist. Before the Revolution, he traveled from New England all the way to the deep South, recounting his adventures in detail and providing a guide to inns and restaurants that the Michelin people would envy today. In Europe, he explored England, France, Italy, and the Low Countries, always associating with the highest-ranking people he could find. His detailed memoirs leave no hint as to how he paid his expenses on these wide-ranging experiences, but he surely got around. Most important, Watson's European travels ignited a lifelong obsession with the marvels of canals. He was especially impressed by the waterway transportation system of the Low Countries, enjoyed a guided tour of the "stupendous works" of the Bridgewater Canal in England, and was knowledgeable about the Canal du Midi in France.[9]

Watson's enthusiasm for canals led to the most spectacular achievement in his lifelong mission of cuddling up to the most famous people he could find. In January 1785, he spent two days at Mount Vernon as George Washington's personal guest. Watson's report of this visit provides a vivid description of Washington at home: "his cautious reserve . . . his urbanity so peculiarly combined in the character of a soldier and eminent private gentleman . . . his smile, which seemed to illuminate his eye; his whole countenance beamed with intelligence, while it commanded confidence and respect. . . . I found him kind and benignant in the domestic circle, revered and beloved by all around him, agreeably social without ostentation; delighting in anecdote and adventure without assumption; his domestic arrangements harmonious and systematic."[10]

Although Watson was suffering from a severe cold and bad cough while at Mount Vernon, he declined Washington's offers of remedies. Washington, however, was resolved to ease his guest's discomforts. As Watson describes what happened after he had retired, "the door of my room was gently opened, and on drawing my bed curtains, to my utter astonishment, I beheld Washington himself, standing at my bed-side, with a bowl of hot tea in his hand."

In the course of conversation with Watson, Washington elaborated on his plans for opening the navigation of the Potomac with canals and locks. Indeed, Watson noted the depth of Washington's concern over this problem and his determination to resolve it. Watson was in enthusiastic agreement with Washington's plan: he had explored the Potomac valley in 1778 and concluded that only the Hudson could compare with it "in magnificence and utility."[11] Washington went so far as to open up his notes on these plans and even to press Watson to settle on the banks of the Potomac. This invitation Watson declined. He went on, however, to lecture Washington on the importance of connecting the various waters of America by means of canals, a conviction he had confirmed by his many voyages on European canals. Washington must have responded eloquently and at length, for Watson reported, "Hearing little else for two days from the persuasive tongue of this great man, I confess completely infected me with the canal mania and enkindled all my enthusiasm."

For the rest of his life, Watson would assert that he was the first person to lay out the concept of the Erie Canal. He claimed he was assured by the knowledgeable people he encountered on a trip over part of the territory in the fall of 1788 that "no such idea existed at that time, otherwise than as we now contemplate balloon stages; or, what is more plausible, a great canal to unite the Pacific with the Atlantic Ocean, at the isthmus of Darien."* His only concession was to admit the possibility that the idea might have occurred to "the enlarged mind" of Henry Hudson himself, had Hudson had a map of New York's inland waters.[12]

* * *

The journey to New York in 1788 originated from a business trip to western Massachusetts, after which Watson decided to continue heading west to indulge in some sightseeing. He crossed over to Albany and

*In 1778, Watson had predicted a population in the United States of 100 million by 1900, but the actual result was 76 million—still, this was a striking projection to have been made only two years after the Battle of Bunker Hill.

headed up the Mohawk valley. The scenery was magnificent, with "the majestic appearance of the adjacent mountains, the state of advanced agriculture . . . and the rich fragancy [*sic*] of the air, redolent with the perfume of the clover, all combined to present a scene . . . [I was] not prepared to witness."[13] He was also impressed by the large numbers of settlers pouring in from "the Connecticut hive, with its annual swarms of industrious and enterprising immigrants, so highly qualified to overcome and civilize the wilderness."[14]

There were problems, however, not the least of which was the lack of comfortable places for a night's rest or a good meal. Watson also describes the roads as almost impassable, cluttered by broken bridges, logs, and stumps, to say nothing of mud so thick his horse sank knee-deep into the road at every step. At one point he encountered a band of Indians "*drunk as lords*," looking like "so many evil spirits broken loose from Pandemonium . . . wild, frantic, almost naked, and frightfully painted."[15]

When Watson reached the headwaters of the Mohawk and the marshy land of Wood Creek, he contemplated the likelihood that "*a canal communication will be opened, sooner or later, between the great lakes and the Hudson*."[16] The imposing prospect led to a further prediction, also to be realized, that the citizens of "the state of New-York have it within their power, by a grand stroke of policy, to divert the future trade of Lake Ontario and the great lakes above, from Alexandria and Quebec, to Albany and New-York."[17]

* * *

Bad weather prevented Watson from continuing westward beyond Wood Creek, but his enthusiasm for the idea of a canal to the west was only beginning to flower. A year later, he moved from Rhode Island to Albany, in order to be closer to where he was convinced the action would soon begin to take place.

He described Albany at that moment as a most unattractive place, a hilly town with streets lacking pavements and ungraded, and no street lamps. The houses were equipped with long spouts that on rainy days poured water on the heads of unwary passersby. He was later to be known by some of the locals as "that infernal paving Yankee."[18] But he knew what he was about, for he was going to play a major role in converting this primitive village into a thriving and cosmopolitan center of commerce and trade.

CHAPTER 5

"A CANAL TO THE MOON"

Elkanah Watson had seen enough to convince himself that a canal across New York State was feasible and would work economic miracles. But his irrepressible enthusiasm and impeccable contacts with the high and mighty were not enough to get the job done. The time had come to orchestrate political support, and that in turn meant preparing documentary evidence and credible witnesses to support his case.

In September 1791 he gathered a small contingent of influential friends and a staff of servants to accompany him from Albany on a six-week trip up the Mohawk and beyond, traveling wherever possible by water, on rivers and across lakes. The most important of his companions was Jeremiah Van Rensselaer, one of the largest landowners in the state and a resolute supporter of Watson's plans. The terminus of the tour was at Geneva, which is situated on Seneca Lake, about ninety miles beyond Watson's stopping point three years earlier at the headwaters of the Mohawk. At that point, having started from Albany, the group had covered about half of what would one day be the route of the Erie Canal.

Although some of the towns the travelers passed through had inns and occasionally churches and law courts, the countryside was more populated with fur-bearing creatures, fish, and eagles than with human beings.* The first national census of 1790 reported the total population of the state as only 340,000 people, including New York City, which worked out to a density of only 6¼ individuals per square mile; even this, however, was more than double the estimates for 1771 under the colonial regime.[1] Nevertheless, more than a quarter of a century after Watson's escorted tour, no town in western New York had a population greater than 6000 and most were less than 3000. Just a tenth of the eight million acres of land along the future route of the canal was under cultivation.[2] We can only wonder how this sequence of sightseers, from Colden to Watson's group, could pass through such a wilderness and still visualize how an extended waterway could create the tremendous economic development that lay ahead.

Watson set to work drafting a report of the trip as soon as he returned to Albany, explaining in detail everything they had seen and done. There was an extended catalog of the landscape, the attractive climate, the economic activity of the areas they visited, and estimates of the work needed to improve the navigability of the waterways they had traveled over.[3] The group had been especially impressed by the abundant output of the farmlands and by the large salt deposits lying just to the north of the Finger Lake region. They expected both to make good use of improved waterway routes to the New York City area and to generate substantial tolls toward making the improvements on these waterways pay for themselves.

Watson summed up his case by asserting that the more he and his companions observed on their way west, the more convinced they became of the necessity to "assist nature" so that loaded boats could travel from the Hudson to the borders of New York State without inter-

*Watson's group spent their first night just west of Schenectady at an inn known as the Mabee House. This tiny structure and neighboring enormous barn were built in the seventeenth century and remained in the same family until the 1970s, when the property was donated to New York State as a most unusual and interesting tourist attraction.

ruption. Then he proceeded to turn up the volume: "The first impression will not fail to be heightened into a degree of enthusiasm, bordering on infatuation."[4]

Watson was not a man to mince words. Nevertheless, there is some question as to how far his imagination reached. A careful reading of what he had to say suggests that his vision of a canal stretched only from Albany to Utica, about a third of the total distance out to Lake Erie. In later years, when the canal was already under construction and Watson was busy defending his view of himself as the first to propose a canal across New York State, he made a revealing confession in the course of recalling his trip west in 1791: "The utmost stretch of our views, was to follow the track of Nature's canal [the rivers and lakes] and to remove natural or artificial obstructions; but we never entertained the most distant conception of a canal from Lake Erie to the Hudson. We should not have considered it much more extravagant to have suggested the possibility of a canal to the moon."[5] This sentence floats alone in all of his vast writings, with no hint of it before he revealed it and no mention of it in any subsequent work.

* * *

As soon as he had finished drafting his report, Watson arranged to meet with his Albany friend and neighbor State Senator Philip Schuyler at their favorite local tavern. By then, Schuyler was already a convert to the wonders of canals, having visited England in 1761 in the midst of the roaring canal boom unleashed by the success of the Bridgewater Canal. Now he was eager to help Watson in getting some action on the canal front in New York. As Schuyler was a war hero, a wealthy property owner, Alexander Hamilton's father-in-law, and a member of an old Dutch family, he was the ideal member of the rich and famous to join Watson's team.

Three days after receiving Watson's report, Schuyler told Watson to prepare a second copy for a Mr. Lush, a member of the legislature, assuring Watson, with much enthusiasm, that he would do everything possible to get a canal law passed during that winter's legislative session.[6] Watson was prepared to go further, however. He submitted lengthy articles on the subject to the *Journal and Patriotic Register* in New York, the *Albany Gazette*, and, anonymously, the *Albany Northern Centinel.*

To top it all off, Schuyler arranged for Watson to send a long document to the legislature in December 1791, which Watson signed as "A CITIZEN." Here Watson proclaimed that the configuration of the land and waters from the mouth of the Hudson north to the branches of the Mohawk and from thence to the "utmost limits of this state . . . are disposed by the *Great Architect of the Universe*, just as we would wish them." Indeed, by merely opening a short canal between the headwaters of the Mohawk and the neighboring Wood Creek—over the Great Carrying Place or Wood Creek Carry—New York State could open a water communication from the Atlantic that would be the most extensive in the world. "Suffice it to say," he concluded, "the [result] would be more precious than if we had encompassed the [Bolivian gold] mines of Potosí."[7]

Watson's eloquence was mighty, but the Great Architect of the Universe had been nowhere near as cooperative as Watson would have wanted his audience to believe. In 1804, Timothy Dwight, president of Yale College and an intrepid traveler through the northeastern United States, declared that navigation on the Mohawk was "so imperfect merchants often choose to transport their commodities along its banks in wagons."[8] And a report of 1818 from De Witt Clinton and the authorities in charge of building the Erie Canal would describe the Mohawk as a "serpentine route" where the entire waterway "became a portage" in the dry season.[9] They would build an artificial waterway over the entire distance from Albany to Buffalo.

Yet, from the vantage point of 1791, improving the navigation of the Mohawk appeared as a compelling project because the payoff could be so enormous. Merchandise as well as humans traveling west from Albany had to begin by wagon instead of boat because no boat could mount the steep slopes to Schenectady, twenty-four miles away; this segment would require twenty-seven locks when the Erie Canal was built. The passage from Schenectady to Little Falls, sixty-two miles up the river, was manageable for small boats of limited capacity, although more than fifty shallow rapids, or "rifts," would provide formidable obstacles. At Little Falls, another portage of over a mile was necessary as crews and passengers had to carry all the gear and all the cargo around the falls and over the hilly countryside. Beyond Little Falls, the forty miles westward to Rome involved another twenty-two sets of rapids. And after that, the westernmost reaches of the Mohawk turned into little more than a mushy creek.

Watson's ringing phrases, accurate or not, seduced a willing audience. A navigable river all the way from Schenectady, or even Albany, to Rome would change the whole course of the nation. The legislature promptly passed the Mohawk Improvement Bill, which had been drafted and sponsored jointly by Senator Schuyler and Governor George Clinton, accompanied by the governor's recommendation that the company established by the act be provided with "every fostering aid and patronage."[10] A sequence of bills established the Western Inland Lock Navigation Company to develop the route up the Mohawk and beyond to Lake Ontario and, if possible, as far west as the Finger Lakes.[11] Another piece of legislation established a Northern Inland Lock Navigation Company to create a navigable waterway due north from the Hudson to Lake Champlain on the border of northern New York State and Vermont.* Senator Schuyler was named president of both companies; the Western Company board of directors included Elkanah Watson and Thomas Eddy, a wealthy Quaker merchant and insurance agent deeply involved in social movements such as prison reform and public welfare.

*One of the surveyors for the Northern Inland Lock Navigation Company was Marc Isambard Brunel, later to become world famous for designing a tunnel under the Thames in London.

* * *

Neither the Western Company nor the Northern Company considered its goal to be canal building. The word "navigation" in these contexts had long meant improving the navigability of the rivers and nothing more than that. Locks to circumvent major obstacles like waterfalls and rapids might be included, but only as devices to bypass rapids, falls, and long stretches of shallow water, with the river itself always as the primary route. This was essentially the structure of Washington's Patowmack Canal Company, even though it contained the mighty word "canal" in its title. Nothing in the Mohawk Improvement Bill and its successors included any possibility of a point-to-point waterway fed by the river waters but otherwise completely separate and apart from them.

Each of the companies was authorized to take subscriptions for a total of one thousand shares at $25 a share, in addition to receiving a grant of $12,500 from the State.* The provisions for the shareholders provided for payments spread over time, as is customary in today's venture capital investing: "The directors of the incorporation shall, from time to time, as occasion may require, call on the subscribers for additional moneys to prosecute the work to effect," but then they added, "whence the whole sum for each share is left indefinite."[12] Not many venture capitalists today would accept that open-ended provision. On the bright side, the legislation provided for an annual dividend of 6 percent of the capital together with authority for the company to raise tolls until the profits were sufficient to cover that dividend. After attempting to guarantee some minimum amount of profitability, the provisions also set the dividend's upper level at 15 percent.

Many of the subscribers were merchants, businessmen, and bankers from New York City. But fifteen out of the original thirty-six had lands along the Western Company's route and stood to benefit from the proj-

*The New York State Museum Web site on the Internet has designs of these shares, providing rare glimpses of the boats in use as well as the construction of the locks at the time.

ect. Their participation was significant. Unlike the great landowners in Britain, who looked down their noses at "industry" while clinging to their pastoral enterprises, American landed aristocrats had a taste for the adventure of commercial enterprises.[13]

Despite the bright expectations, these investors were in no hurry to deliver the hard cash to which they had committed themselves. In an effort to galvanize more support, Schuyler announced in May 1792 that he would increase his own subscription from ten shares to one hundred. Schuyler also enlisted the participation of the famed financier and land speculator Robert Morris (no relation to Gouverneur) whose personal credit had financed a significant portion of the American Revolution. Morris had also served with distinction as a delegate to the Constitutional Convention in 1787 and had been a member of the first U.S. Senate.

There was never enough money, even with this high-powered backing. In the end, 743 out of the authorized 1000 shares were sold, but 240 of those were forfeited because their owners refused to meet the calls for additional capital. By 1801, the stockholders had been called nine times to increase their investments. The State had to donate an additional $10,000 in 1795 and then lend the company another $37,500 the following year. Total outlays for construction amounted to $400,000, but the investment was never a rewarding one. The company paid only two dividends in its lifetime, one of 3 percent in 1798 and another in 1813.

* * *

The novelty of the project and the lack of skilled personnel created problems from the start. Poor Schuyler had to appoint himself chief engineer, although he was suffering from gout and was still active in the State Senate.[14] De Witt Clinton, one of Schuyler's political opponents and then secretary to Governor George Clinton, his uncle, took a dim view of Schuyler's decision and castigated Schuyler in a newspaper article as a "mechanic empiric . . . [who is] wasting the property of the stockholders."[15]* Schuyler's feelings may have been hurt by the intended insult—and he was doubtless fully aware of Clinton's political ambitions—but he could hardly have disagreed. As he complained in 1793 to William Weston, a British engineer he hoped to employ, taking on this responsibility was a counsel of despair, for he was "without the least practical experience in the business."[16] Then Schuyler and Watson began to quarrel over Schuyler's salary and Schuyler's "tyrannical manner," bringing to an end both a beautiful friendship and Watson's active participation in the company's affairs.[17] Watson soon moved himself from New York to rural Massachusetts, where he would launch the first livestock and agricultural fairs, which developed into major annual events throughout New England.†

Meanwhile, the clearing of the Mohawk and the building of locks moved ahead despite the obstacles, the interruptions, and the inexperience of the designers, contractors, and work crews. More than four hundred men, recruited from Pennsylvania, Vermont, Connecticut, and Canada, worked on the project at a time, but the labor supply on the Mohawk was as stubborn an obstacle as it had been for Washington on the Potomac. Although the New Yorkers did not follow Washington in reverting to slavery to overcome that difficulty, they did employ Irishmen who emulated their fellow countrymen on the Potomac by engaging in riots against the local population.

The effort of building locks between Albany and Schenectady, which

*At a later date, Clinton was more generous to Schuyler. Still insisting Schuyler was "not a practical engineer," Clinton, writing pseudonymously, went on to say, "Without his talents and services, [the Western Company] would never have been commenced and prosecuted." See Tacitus, *The Canal Policy of the State of New York*, p. 18.

†Watson was convinced that by encouraging competition among farmers, he could make a significant contribution toward improving both the quality and the quantity of New England's agricultural production and stimulate American industry at the same time. The appearance of prizewinning animals and awards to the finest produce at Watson's fairs attracted increasing attention and had a lasting influence on agricultural practice in the northeast United States. On occasion, Watson would display broadcloth made up by "the best artists in the country" and woven from the wool of his fairs' prizewinning sheep. See Watson, *History of the Rise, Progress and Existing Condition . . .*, pp. 10 ff., for a long and interesting account of this ambitious effort to stimulate economic development.

included bypassing the massive Cohoes Falls, was beyond the limited engineering skills available. This stretch would also pose a major challenge thirty years later when the Erie Canal itself was under construction. Work therefore started in a westerly direction from Schenectady, the true gateway to the Mohawk valley. The most significant improvements to navigation on the Mohawk were at Little Falls, sixty-two miles farther west, where the river fell over forty feet in three-quarters of a mile. The company bypassed the falls with five locks over a reach of about a mile, with each lock lifting boats by nine feet. It also built two miles of canal with two locks near Rome to fulfill Watson's dream of a connection between the Mohawk headwaters and Wood Creek, cutting the travel time at this crucial link from a day to an hour and opening the way by water to Oneida Lake and thence up to Lake Ontario.

The locks were built out of wood, which soon leaked and rotted even though cut from the excellent timber growing in profusion in the virgin forests on either side of the Mohawk. Later on, around 1803, the builders used bricks and then stones. Although some of these stone locks were elegantly put together, they lacked a durable mortar to keep them from leaking. After William Weston came on the job in 1795, he developed an effective mortar that managed to bind the locks in more permanent fashion—but leaks never ceased to be a problem.

The tolls, although high enough to provoke complaints from users of the waterway, were never sufficient to finance expansion and improvement; maintenance swallowed up just about all the revenue. The company increased the tolls in the summer, when the water flow of the Mohawk was low, but the traffic flow then diminished in response. Significant amounts of output from the local farms continued to move eastward by road instead of on the waterway.[18]

In 1803, Schuyler left the Western Company to serve as U.S. senator from New York. As an odd coincidence, he was going to fill out the term of De Witt Clinton, who had decided he would prefer to be mayor of New York City than stay mired in the primitive living conditions and political swamp of Washington, D.C.

* * *

Despite all its shortcomings, the Western Inland Lock Navigation Company was able to accomplish a great deal. It was, after all, the first meaningful effort to improve water navigation to the west since "the neck digged through in 1730." The results are all the more remarkable because they were achieved at a time when civil engineering was a non-existent profession in the United States, and only a few individuals like Schuyler and Watson had studied as well as visited the canals of England and France. Almost everything was done on an ad hoc basis, with hardly any plans drawn out on paper. These informal methods of "American ingenuity" continued to provide solutions to the most complex problems even when the Erie Canal was under construction.

In addition to constructing the locks, the company substantially improved the navigability of the Mohawk by clearing out sandbars, stones, and sunken timber and by erecting a series of small dams to increase the water flows on the many rifts that characterized long stretches of the river, especially between Schenectady and Little Falls.

The locks may have been fragile, but they did bypass Little Falls, and the ones farther west linked the Mohawk and Wood Creek by water. Boatmen no longer had to face the backbreaking and dreary business of dragging their boats overland around the falls. And, thanks to the short canal linking the Mohawk to Wood Creek, the Wood Creek Carry became history.

These were no minor achievements. With an uninterrupted waterway available, the boatmen could replace their simple bateaux, carrying less than two tons of cargo, with flatboats called Durham boats that could transport up to twenty tons of iron or 150 barrels of flour on their downstream runs. The Durhams provided dramatic proof of the superiority of waterways over the bumpy land route for transporting heavy cargo or large numbers of people.

These extraordinary vessels, which had played a crucial role in Washington's crossing of the Delaware in December 1776, were named after the engineer Robert Durham, who developed them in Pennsylvania in 1757. With flat bottoms, they were often as long as 65 feet and about 8 feet in the beam. Aided by 18-foot oars, the Durhams traveled swiftly downstream and steered easily through the rapids, while their 30-foot masts enabled them to glide along through calmer waters under sail at high speed. On the upstream trips, the Durhams carried a crew of six men and a captain, with two men pushing against the currents with poles and the other four rowing against the current.[19]

The Durhams were the Mack trucks of the era, and the significant improvements to the navigability of the Mohawk carried out by the Western Inland Lock Navigation Company made it possible for the Durhams to negotiate the entire distance from Schenectady to Lake Oneida by water, except when the river froze or flooded. The greater capacity of the Durhams, in turn, enabled the boatmen to reduce the fares they charged for people and for cargo, thereby expanding the volume of westward movement. According to a 1798 report of the Western Company directors, the cost of transportation from Albany to the environs of Geneva, a distance of almost 200 miles, fell from $100 a ton to $32, while the cost from Albany to Niagara Falls was sliced in half.[20] Other papers of the directors comment on the downstream cargoes of meat, flour, furs, lumber, pearl ash (a kind of potash), wheat, butter, lard, and salt; manufactured goods included cotton, linen, and glass. Meanwhile, European and Indian products and manufactured goods moved upstream.[21]

These reductions in cost and the mention of the growing number of products on the Western Company waterways represent profound change that would reverberate for many years to come. As a rule of thumb, the cost of transportation should equal no more than 50 percent of the price charged to the final customer. But at that time, the cost of transporting commodities by wagon over the rough roads from Buffalo to Albany often involved sums equal to five or six times the values of the goods themselves.[22] By so dramatically reducing transportation costs, the waterway constructed by the Western Company greatly widened the variety of commodities that could now come to market, and a nation-wide economy began to bud from the small seeds the company had planted.[23]

Not incidentally, this increase in activity ignited an outburst of speculation in land along the route and even beyond to the farther western reaches of New York State. This was a classic bubble, but it popped in only a short time. Speculative bubbles need a constant inflow of capital to keep driving prices upward, and capital was scarce in the United States of the 1790s. When the bubble burst in 1796, the reaction was so violent it caught even the renowned financier Robert Morris in a mess of confused titles, with taxes and debts unpaid. Morris ended up in debtors' prison. It was a tragic end to the career of a man who had labored to help the young American nation finance the Revolution, had served with distinction as a delegate to the Constitutional Convention in 1787, and was a member of the first U.S. Senate.[24]

* * *

The Western Inland Lock Navigation Company's failure to do much work beyond Wood Creek was not a result of shortsightedness but of money. Watson was continually pressing for an extension beyond Wood Creek at least to the Finger Lakes, but he got nowhere. In 1796, Thomas Eddy, one of the directors and among Watson's companions on the 1791 expedition, teamed up with the English engineer William Weston to study the possibilities of a canal that would avoid Wood Creek and Oneida Lake by cutting right through to the Finger Lakes from the head-waters of the Mohawk. In response, the legislature shelved their report, leaving the few hardy pioneers in the western part of the state in the primeval wilderness, without any transportation system to move their crops to market or to bring in their necessities. Farming remained a home-based function rather than the substantial commercial enterprise it would become when the Erie Canal was finally in operation.

Yet the Western Company was in many ways a proving ground and training school for the great artificial waterway that would one day span the state all the way from the Hudson to Lake Erie. Within eleven years after its charter was issued in 1792, a group of men with only the most limited kinds of engineering experience converted the unstable and treacherous Mohawk River into a continuous deep water channel for large Durham boats as far as the Wood Creek Carry. It is fair to speculate how much acceptance the Erie Canal would have achieved without the experience of the Western Company to build upon.

At one time or another, and either directly or indirectly, many of the leading engineers and politicians involved in the creation of the Erie Canal had been associated with the Western Company. The papers of the directors repeatedly emphasized the importance of settling the western part of the state, the possible development of other routes to Lake Erie, and the benefits to both state and nation that lay ahead. The Western Inland Lock Navigation Company's accomplishments did serve the important purpose of keeping the public's attention directed toward the notion of a navigable waterway through the mountains. All its difficulties and shortcomings only made the designers of the Erie Canal even more certain that an artificial waterway would be superior to a combination of canal and navigable river waters.

The financial sorrows of the company provided the most important lesson for the future. Jefferson may have opined to Washington that "public undertakings are carelessly managed, and much money spent to little purpose," but the experience of the Western Company made it clear that ventures this large would overwhelm the limited resources of private investors in the United States. Other and radically different forms of finance, including public finance, would have to be developed.

* * *

Nevertheless, the Western Inland Lock Navigation Company occupied so much of everyone's attention in the 1790s that discussion of cutting through to the west with something even more ambitious subsided into little more than occasional sputters. One of those sputters came from Gouverneur Morris. In the course of a conversation in 1803 with New York State Surveyor General Simeon De Witt (a cousin of De Witt Clinton), Morris mentioned the possibility of "tapping Lake Erie . . . and leading its waters in an artificial river, directly across the country to the Hudson River."[25] De Witt considered Morris's fantasy a "romantic thing, and characteristic of the man."[26] James Geddes, who had started out in life in the salt business but later become a local judge and an assistant in the surveyor general's office, also discussed the matter with Morris. After listening to Morris for longer than he wished, Geddes observed that "it was almost impossible to call his attention to the impracticality of such a thing."[27]

In an odd coincidence, two years later in 1805, Geddes appears to have discussed the idea of a canal from the Great Lakes to the Hudson with a local merchant named Jesse Hawley. Hawley was then a boarder in Geddes's home in Geneva. If Gouverneur Morris and Elkanah Watson were the spiritual grandfathers of the Erie Canal, Hawley would turn out to be its first really hardheaded proponent.

Hawley's son claims that the idea of a canal between the interior and the east occurred to his father in 1805 during a discussion with one of his suppliers over the difficulties and uncertainties of shipping flour east by road. Hawley knew whereof he spoke, for the roads in New York at that time were a disaster. Most roads were the old Indian trails, widened and packed down with dirt where necessary but brutally dusty in dry weather and often swampy in wet weather. Travel through the muck of the spring thaw was next to impossible and then the ruts and potholes of the spring remained to torment summertime travelers as well. In the most marshy areas, the roads were composed of logs set across the route with earth between them in a pretense at smoothing. Although this device kept the road from sinking into the marsh, it earned the nickname of "corduroy," and for good reason. To make matters even worse, many streams lacked bridges connecting a road from one side to the other, so that cargo had to be forded across rushing waters.[28] These hurdles were not about to disappear. In 1816, a traveler through western New York would describe the main road as "a causeway formed of trunks of trees . . . we could by our feelings have counted every tree we jolted over."[29]

Hawley's supplier agreed enthusiastically with Hawley's suggestion of a canal across the state but then pointed out that a waterway like the one Hawley envisioned would be impossible without a large head of water to keep it filled. When Geddes raised the subject with Hawley over the dinner table one evening shortly after that conversation, Hawley grabbed Geddes's map, spread it out on the table, and, as he later described the moment, "ruminating over it, for—I cannot tell how long—muttering *a head of water*; at length my eye lit on the falls of Niagara which instantly presented the idea that Lake Erie was *that head of water*."[30] Hawley was hooked. He soon had no other topic of conversation, but when he mentioned his suggestion of connecting the Hudson to Lake Erie "*by a canal!* [I was] generally laughed at for my whim!"[31] He was not to be deterred by ridicule.

Hawley was in many ways a most unlikely individual to have played such an historic role. His education never went beyond a country school: he was neither an engineer nor a surveyor. He was not even a high-class dreamer like Morris.

Hawley earned his living in the humdrum business of forwarding flour to New York City from mills in the interior near where Rochester would one day thrive. This trade made him a good and steady customer of the Western Company, despite his constant complaints about how the higher tolls of the summer season cut deeply into his company's profits and pushed him and his partner into debt. The debts grew so heavy that he and his partner finally fled to Pittsburgh to elude their creditors. Even flight would not cool Hawley's ardor for the notion of a canal. Once settled in Pittsburgh, he went into print about his obsession for the first time, submitting an article to a local newspaper. But then Hawley's conscience got the better of him. He returned to New York, surrendered to the authorities in August 1807, and spent the next twenty months in debtors' prison in Canandaigua on the Finger Lake of the same name.

Incarceration was hardly a waste of time. After recovering from "despondency at the thought that hitherto I had lived to no useful purpose," he resolved to "publish to the world my favorite, fanciful project of an overland canal, for the benefit of the country, and endure the temporary odium it would incur."[32] Only three months after he had entered prison, the first of Hawley's fourteen essays on the subject appeared in the *Genesee Messenger*, a paper published in Canandaigua; the series ran all the way to April 1808. Hawley used the pen name of Hercules, an appropriate choice for the tone of his work.

A sense of destiny and greatness pervades these essays. Consider the following: "The trade of almost all the lakes in North America, the most of which flowing through the canal, would centre at New-York for their common mart. This port, already of the first commercial consequence in the United States, would shortly after, be left without a competition in trade, except by that of New-Orleans. In a century its island would be covered with the buildings and population of its city."[33]

Although Hawley grumbled that the public treated his work as "the effusions of a maniac," these lengthy essays are extraordinary for the way they combined the boldness of vision with his painstaking attention to an immense compendium of detail. He goes on at length about methods of construction and provision of water supplies, the most minute parts of the route such as the shifting altitude from sea level as well as distances, the powerful buildup of arguments as to why the national government should finance the venture, an estimate of the total cost at $6 million (strikingly close to the actual figure), and helpful comparisons with British canals and the Canal du Midi, to mention just a few of the matters that he subjects to his tireless analysis.

How did Hawley manage the task he set for himself with such

superb skill, given his background and the limitations under which he prepared these essays? Although he may have had willing and enthusiastic research assistants among the guards and other employees of the Canandaigua prison, they were hardly likely to have been scholars or surveyors. The true source of much of his amazing achievement remains a mystery.

Hawley's fecund imagination is not satisfied with projecting only a canal across New York state. In Essay XIII, titled "Other Improvements Proposed," he explores the many possibilities of developing the waterways in other states, including Virginia (he approved of Washington's efforts), South Carolina, Georgia, Ohio, and Tennessee, as well as the Mississippi River itself. He follows this set of recommendations with a remarkable prediction: "A marine canal, the most noble work of the kind on this 'ball of earth,' would be a cut across the Isthmus of Darien. Were the Mexican empire under an independent government—or even under an enterprising one—this would be done in less than half a century, and those provinces opened to a liberal trade, under which their abundant resources would make them immensely wealthy."

Hawley concludes his essay with a comment on the importance of freedom for human advancement that is as relevant to today's world as it was in 1807: "Nature never has, nor will, endure the jealousy and selfish dogmas of man with impunity. From the huckster's shop to the chartered company's shipping warehouse, the principle continues the same. Wherever the avarice and vanity of man has imposed his restrictions—whether in religion, politics, or commerce—she has entered her caveat to them."

Hawley typifies the view of life and the Almighty held by Americans in the early years of the nineteenth century, some of which lingers on into our own time. He did not dodge the consequences of his acts. No one was promised a rose garden. Nevertheless, God has provided Americans with the setting, the essential tools, and, above all, the intelligence and drive to accomplish whatever the appointed task may be (a gift the early Americans were certain had been denied to the natives of their land). Thus:

> [When] we turn our reflections to the fatigue and toil of so much land transport, we are apt to exclaim—Why was not the parent of nature so thoughtful—why was he not so kind, as to give this country a river navigation from the Atlantic to the lakes, like that to Albany? Why these murmurs? The Creator has done what we can reasonably ask of him. By the Falls of Niagara he has given a head to the waters of Lake Erie sufficient to flow into the Atlantic by the channels of the Mohawk and the Hudson, as well as by that of the St. Lawrence. He has only left the finishing stroke to be applied by the hand of art, and it is complete! Who can reasonably complain? . . . If the project be but a feasible one, no situation on the globe offers such extensive and numerous advantages to inland navigation by a canal, as this! . . . It would be a burlesque on civilization and the useful arts, for the inventive and enterprising genius of European Americans, with their large bodies and streams of fresh water for inland navigation, to be contented with navigating farm brooks in bark canoes.[34]

Jesse Hawley, bankrupt businessman and jailbird, accomplished more than anyone up to that point in provoking action to build an uninterrupted waterway across the State of New York. His eloquence, analytical abilities, and the massive amount of critically significant information in his essays attracted attention in the highest places. Later the essays would serve as a kind of guidebook for the series of commissions sent westward to arrive at definitive plans, budgets, and supply lines for the canal that many commentators would repeatedly call a "stupendous" achievement. Indeed, almost immediately after the appearance of Hawley's essays, deeds finally began to replace words.

PART II

The Action Begins

CHAPTER 6

THE SUBLIME SPECTACLE

Nothing will motivate a politician faster than money. In his second inaugural address of March 4, 1805, President Thomas Jefferson dangled the money. First, he proudly announced the nation's steadily expanding budget surplus and the early prospect of redeeming the national debt right down to the last dollar. Receipts in the year 1804 had amounted to $11.8 million versus a mere $4.4 million in 1790; expenditures were $8.7 million versus $4.3 million.[1]

And then he proposed that "the revenue thus liberated may, by a just repartition of [the surplus] among the States and a corresponding amendment of the Constitution . . . be applied in time of peace to rivers, canals, roads, arts, manufactures, education, and other great objects within each State." These noble goals were embodied in Jefferson's dream of what a great republican government owed its constituents, but his words were electrifying to members of Congress, state legislatures, and governors. But more good news was to come. In December 1806, Jefferson reported that the swelling surplus was running well ahead of estimates. And in his annual message to Congress in early 1807, he once again recommended spending the surplus on "internal improvements," although on this occasion he spoke in generalities rather than specifying the projects he had suggested in 1805.

Jefferson reminds his audience, "The suppression of unnecessary offices, of useless establishments and expenses, enabled us to discontinue our internal taxes. These, covering our land with officers and opening our doors to their intrusions, had already begun that process of domiciliary vexation which once entered is scarcely to be restrained from reaching successively every article of property and produce."

Consequently, he continues, "It may be the pleasure and the pride of an American to ask, What farmer, what mechanic, what laborer ever sees a taxgatherer of the United States?" Up until the enactment of an income tax during the Civil War, almost all the federal government's revenue took the form of tariffs levied on imported foreign merchandise and collected at the ports and frontiers. The IRS of the day was little more than a customs-collecting service.

To Jefferson and his fellow citizens, spending surplus revenue at home appeared far preferable to relieving tax burdens on foreigners by reducing the tariffs. Only those few individuals who had read Adam Smith's *The Wealth of Nations*, published in 1776, would understand that import taxes are paid for the most part by the citizens of the importing country in the form of higher prices, even if, as Jefferson concedes, the duties are "paid chiefly by those who can afford to add foreign luxuries to domestic comforts." Jefferson to the contrary, the customs agents were indeed tax gatherers in the fullest sense of the word.

Politicians hastened to respond to Jefferson's offers to share the wealth. On March 2, 1807, the Senate authorized the secretary of the treasury to prepare "a plan for the application of such means as are within the power of Congress, to the purposes of opening roads, and making canals . . . which as objects of public improvement, may require and deserve the aid of government."[2] Thirteen months later, Albert Gallatin, the Swiss-born secretary of the treasury, respectfully submitted his recommendations, with apologies for unavoidable delays but he did point out that selecting, arranging, and condensing his material was "a work of some labor."[3]

Indeed it was. But Gallatin was no stranger to either hard work or controversy. In 1780, at the age of nineteen, he gave up a fortune and high social position to immigrate because of his "love for independence in the freest country of the universe." He taught French at Harvard briefly after arriving in the United States and finally settled in Pennsylvania. As a member of the House of Representatives, his financial skills were already so apparent that he was made a member of the first standing committee on finance, subsequently to be known as the Ways and Means Committee. Gallatin's obsession with meticulous bookkeeping led to the establishment of accounting practices at the Treasury that are still in use today. After the War of 1812, he negotiated a trade agreement with Great Britain that abolished discriminating duties. After serving as minister to both France and Britain, in 1817 he became the president of a major New York City bank and lived to the ripe old age of eighty-eight.

With his long, solemn face and protruding nose, Gallatin had been constantly in conflict with Washington's secretary of the treasury Alexander Hamilton. It was Gallatin who first proposed the law requiring the secretary to submit a report of his department to Congress and then was instrumental in establishing the House Ways and Means Committee to assure the Treasury's accountability to Congress. But when he became secretary himself in 1801, in Thomas Jefferson's first administration, Gallatin followed in Hamilton's path, making every effort to keep the Treasury independent of legislative interference. He was such an effective Cabinet member that he stayed on the job through both of Jefferson's terms and into James Madison's second, for a total of thirteen years—the longest tenure in history.

* * *

Gallatin's report of 1808 is a remarkable document, masterly in its presentation, keen in its intuitive grasp of the power of networking, authoritative in its handling of the economics of internal improvement, and fascinating in its wealth of detail about the development of transportation systems. His work could well serve as a textbook for the basic principles of economic development in our own time.

Gallatin rests his case on the principle that improved transportation is an indispensable condition for increasing national income and wealth. Good roads and canals do more than just reduce the expenses of moving cargo and people. An improved transportation system would make it possible for the first time to move goods that had been excluded from trade and commerce, due to their excessive weight or the distance of their origin from potential markets. By increasing the volume of goods entering trade and commerce, the transportation system would instantaneously increase national wealth.

Quite aside from Gallatin's keen insights into the rewards to the nation of these internal improvements, he had a strategy to achieve his goal in the shortest possible time and with the least political infighting. Each project would matter to the communities immediately affected more than to communities located farther away, but by combining the interests of all in advance in one giant piece of legislation, the entire United States would be engaged in creating a single great system instead of a fragmented process in which the citizens of each state would be trying to get their hands on the money before some other state got there first.*

By that time, the inland territories of Vermont, Kentucky, Tennessee, and Ohio had joined the original thirteen states. Most localities in the early 1800s, like these new members of the union, had little interest in trading with their neighbors, who, except along the Atlantic seacoast, were far away from one another or were producing the same types of goods and services. Consequently, Gallatin explains, the first pathways of communication built in an undeveloped economy like the young United States were destined to be unproductive. In order to do any volume of business, connections over long distances were absolutely essential. The

*For an extended and stimulating survey of these developments, see Henry Adams, *The Life of Albert Gallatin*, pp. 350–55.

urgent need, Gallatin explained, was for a nationwide network of roadways and canals linking each local pathway to the larger whole.

Gallatin complains that progress in these directions in the United States has been slow, because the resources of private investors or local governments are too small to achieve this greater goal. As he points out, "The great demand for capital in the United States, and the extent of territory compared with the population . . . are, it is believed, the true causes which prevent new undertakings, and render those already accomplished [such as the Western Inland Lock Navigation Company], less profitable than had been expected."[4] And so, Gallatin concludes, "The general government can alone remove these obstacles. With resources amply sufficient for the completion of every practical improvement, it will always supply the capital wanted for any work which it may undertake."[5]

Gallatin emphasizes an additional compelling consideration. An efficient nationwide transportation network will stimulate a community of interests that would unite people in every part of the nation. Nothing else within the power of government could do more to reinforce "that union, which secures external independence, domestic peace, and internal liberty."[6] Here he echoes Washington's pivot and anticipates De Witt Clinton's eagerness to establish a grand transportation system in New York, but he reflects the thinking of an even more auspicious authority, Adam Smith, who declared in 1766, "The division of labour, in order to opulence, becomes always more perfect by the easy method of conveyance in a country. . . . Water carriage is another convenience, as by it 300 ton can be conveyed at the expence [*sic*] of the wear and tear of the vessel, and the wages of five or six men, and that ton in a shorter time than by a hundred wagons which will take six horses and a man each."[7]

Gallatin does mention canals between the Hudson and Lake Erie, which he considered of first-rate importance, but his scheme differed significantly from what Hawley had been considering. He describes two canals, one running beside the Mohawk and from there up to Lake

Ontario and a second canal connecting the western end of Lake Ontario to Lake Erie and bypassing Niagara Falls. His estimate for this project is in the area of $3 million, about half the ultimate cost of the full 363-mile canal running all the way from Albany on the Hudson to Buffalo on Lake Erie.

This is just one example of Gallatin's detailed analysis of canals and roadways throughout the country. But Gallatin also includes contributions from two famous experts, Robert Fulton, who needs no introduction, and Benjamin Latrobe, the great Yorkshire-born architect who introduced the Greek revival style to the United States and was the primary architect of the Capitol in Washington.

Latrobe takes a dim view of canals like the Patowmack, designed only to bypass waterfalls and other difficult passages of rivers. As Benjamin Franklin had written in 1772, in a letter from London to the mayor of Philadelphia, "Rivers are ungovernable things, especially in Hilly countries. Canals are quiet and very manageable."[8] Franklin was no doubt aware of the even stronger statement on this matter by James Brindley, the engineer of the Bridgewater Canal, who once said that small rivers "were intended by the Almighty for feeding canals."[9] These were precisely Latrobe's views.*

It is interesting to note that the Spanish clergy in the 1580s had taken precisely the opposite position, declaring, "If God had intended for the rivers to be connected, He would have made them so."[10] In view of the tremendous importance of efficient transportation networks and the poor navigability of her natural rivers, Spain's failure to develop economically in step with the rest of western Europe was a significant but unfortunate consequence of this attitude.

*Latrobe's charming two-page "Postcript" [*sic*] on the attractions of "rail roads" (pulled by horses), written in 1808, is worth the price of admission. Latrobe sees only a limited future for this device, however, because "the sort of produce which is carried to our markets is collected from such scattered points, and comes by such a diversity of routes, that rail roads are out of the question as to carriage of common articles (Gallatin, p. 107)."

Robert Fulton had only recently returned to the United States from an extended stay in Europe. He was both showman and genius, combining many of the talents of P. T. Barnum, Thomas Edison, Bill Gates, and Leonardo da Vinci. He had originally crossed the Atlantic to improve his skills as a painter of miniature portraits, his primary source of income at the time. His refined facial structure and gracious manners gave him the air of an artist all his life, but his natural engineering and inventive talents soon overtook his interest in becoming a famous painter.* He had already invented a submarine, in which Napoleon had shown great interest. In 1796 Fulton published a book on small canals, which he boasted about in correspondence to George Washington a year later.

Now Fulton was fixated on the notion of canals for bringing great quantities of merchandise to market at prices far cheaper than road transportation. His analysis in the Gallatin report is a dazzling performance in support of this viewpoint. He provides a wide variety of examples from existing routes in the United States and reveals a keen sense of how such reductions in the cost of moving freight can enhance and enrich the process of economic growth.

Fulton estimates that construction of a canal would cost about $15,000 a mile, a figure remarkably close to the $6 million spent on building the Erie Canal, where work would not even begin for another nine years.

The primary attraction of moving cargo on a canal instead of on a river is in the calm and flat surface over which a canal boat can travel without having to deal with either upstream or downstream currents. A well-built boat floating on water could carry much more freight at no slower a speed than a horse-drawn wagon rumbling along a bumpy road. On most canals at that time, boats had no motive power of their own but were pulled along by horses or mules walking on a towpath by

*Not quite. Fulton's engineering drawings are small masterpieces.

the side of the canal, with a boy leading or riding on the horse and a man on board steering the boat. Although this sounds like a poky, primitive means of locomotion, in reality it was the critical technological advantage over travel by road or by river. By allowing the boat to move smoothly along its waters, the canal avoided all the heavy human efforts spent poling boats upstream on rivers and controlling the speed when moving downstream—major obstacles on both the Potomac and the Mohawk. As a result, a canal could carry larger boats, with more cargo-carrying capacity than boats forced to confront the trials of river travel.

Fulton begins his argument with an actual example involving a canal boat with one man, one boy, and one horse, the usual combination. This boat is moving 25 tons of cargo twenty miles a day. Fulton estimates that the man, the boy, and the horse would cost a total of $2.50 a day. Tolls to cover the cost of maintaining the canals, plus tolls at locks, inclined planes, tunnels, and aqueducts would add another $2 a day. He completes his estimate with 50¢ for the interest on the wear of the boat. The total comes to $5 a day for moving 25 tons twenty miles, which works out to 20¢ a ton, or, more simply as a rule of thumb, $1 a ton for a hundred miles. The cost by road, over the same distance, he points out, would be $10 a ton.[11] Even on a road of the best kind, Fulton explains, four and occasionally five horses are often required to transport as little as 3 tons, or less than 1 ton per horse, while one horse can draw 25 tons on a canal. The saving that results is not only in the cost of the horses themselves, but their feeding, shoeing, gear, wagons, and care.[12]

Along the way, Fulton calculates, "The merchandize which can bear the expense of carriage on our present roads to . . . any . . . distance of 300 miles, and which for that distance pays 100 dollars a ton, could be boated on canals *ten thousand miles for that sum*."[13] No wonder, then, that he comes out loud and clear in favor of canals, and not just for the dramatic reductions in transport costs they provide. Eloquent and persuasive as Fulton is, he fails to mention the most compelling element of his whole argument: that by reducing the cost of transport from $10 a ton to $1 a ton, it would be profitable to move a far greater variety and quantity of goods to market. This is the very essence of what economic growth and development are all about.[14]

There are also the vast benefits of networking: "When the United States shall be bound together by canals, by cheap and easy access to market in all directions, by a sense of mutual interests arising from mutual intercourse and mingled commerce; it will no [longer be] possible to split them into them into independent and separate governments, each lining its frontiers with fortifications and troops, to shackle their own exports and imports to and from the neighboring states."[15] Fulton ends with a charming quotation from his book on small canals, in which he contemplated the time when "canals should pass through every vale, wind round each hill and bind the whole country together in the bonds of social intercourse."[16]

* * *

All of these writings, debates, speeches, and proposals were symptomatic of a more basic set of trends. The economy of the young United States was developing so rapidly that the rudimentary transportation system of the late 1700s was not just obsolete—it was a positive restraint on economic growth.

By the time of Jefferson's inaugural address in 1805, the nation's population had grown to 6.3 million people from 3.9 million in 1790, an annual growth rate of 3.2 percent (population in the 1990s grew at an annual rate of about 1 percent). Merchandise exports had quintupled over the same time period, surpassing $100 million in 1805. Imports had grown even more rapidly. Gross tonnage moving across the seas on American vessels had doubled. Perhaps most revealing, during the years from 1803 to 1807, an average of eighty patents were issued for new inventions annually, four times the rate from 1790 to 1794.[17]

What Cadwallader Colden, Christopher Colles, Gouverneur Morris, Elkanah Watson, and Jesse Hawley had foreseen was just beginning to come true. More than dreams would be involved from this point forward: the need was real. In New York, serious business was stirring.

* * *

The legislature of New York State lost no time in reacting to Jefferson's seductive words to Congress and the arguments and evidence presented in Gallatin's powerful analysis of the crucial role of canals and public roads. On February 4, 1808, even before Gallatin's report had made its appearance, Joshua Forman, an assemblyman from Onondaga County,* introduced a resolution beginning with these phrases: "Whereas, the President of the United States, by his message to Congress . . . did recommend, that the surplus moneys in the treasury . . . be appropriated to the great national objects of opening canals. . . . And whereas, the state of New-York, holding the first commercial rank in the United States, possesses within herself the best route of communication between the Atlantic and western waters. . . ."

According to Forman's own recollection, he was inspired to take this action after a leisurely evening in the company of Benjamin Wright, an assemblyman with whom he shared quarters. Wright, the son of a debt-ridden lieutenant in Washington's army, had had little education but he had developed into a notably accurate surveyor. Among his earlier assignments, he had done most of the surveying on the Mohawk for the Western Inland Lock Navigation Company in the 1790s.[18]

In the course of the evening, Forman and Wright discussed an article on canals in the latest issue of Rees's *Cyclopedia*, an English periodical to which they subscribed. This article provided a detailed description of the development of England's elaborate canal system in the years after the opening of the Bridgewater Canal, concluding from this experience that canals were a far more effective form of transportation than costly efforts to improve the navigability of rivers or to build roadways.

*Onondaga County sits just about halfway between Albany and Buffalo, with Syracuse—first organized as a town by Joshua Forman in 1825—at its center.

Forman immediately recognized from the *Cyclopedia* that the case for canals was a perfect fit for his own political backyard in central New York. Why not build an uninterrupted cross-country canal from the Hudson directly to the Great Lakes, bypassing all the rivers such as the Mohawk and even Lake Ontario? Wright, drawing on his experience along the Mohawk, told Forman the idea would be "a folly." But Forman had a key insight—or foresight—that broke down Wright's resistance: an uninterrupted waterway across the rich country from the Hudson to Lake Erie would readily pay for itself as a line of new towns built up along the waterway would develop and enrich the whole area.

The two men agreed on the immense importance of the project and the urgency of determining its practicality. After spending the night on the task, Forman introduced his resolution in the Assembly the following morning.

Forman was well aware that no one would listen to him if he made his proposal as a work of the State of New York. Therefore, after his preamble of "whereas'es," his resolution recommended the establishment of a joint committee to arrange for exploring and surveying "the most eligible and direct route for a canal to open a communication between the tide waters of the Hudson River and Lake Erie; to the end that congress may be enabled to appropriate such sums as may be necessary to the accomplishment of that great national object." But Forman was in trouble right away. Even though he had invoked the president and Congress, his fellow assemblymen jeered him.

Although upset at this reaction, Forman was so convinced of the strategic importance of his plan that he kept right on speaking, and at length. He emphasized the attraction of transport over a landscape that stretched from the headwaters of the Mohawk as far as Niagara Falls without mountains, and with no large rivers to cross, so that the elaborate and costly tunnels and aqueducts of the European canals could be avoided. Using the experience of the Canal du Midi as a basis, he estimated the total cost of the overland canal at $10 million—which turned out to be a significant overestimate—but which "must appear as a bagatelle to the value of such a navigation." In addition, the rapid settlement of the territory would form a dense barrier toward Canada as well as a prospering outlet for trade and commerce. Finally, following in the footsteps of those who presented a more ambitious vision of the canal, Forman underscored the indissoluble bond of union the canal would create between the western and the Atlantic territories, "chaining them to our destinies in any national convulsion."

Forman carried the day, sort of. The resolution passed the legislature the following April, but, as one member described it, this happened in large part because it "*could do no harm* and *might* do some good." They appropriated $1000 to cover the expenses of the survey proposed by the resolution, then thought better of it and cut the amount down to just $600—a paltry sum even in those days. Nevertheless, a subsequent resolution directed the surveyor general to have the survey carried out, with maps, for submission to the president of the United States.

* * *

State Surveyor General Simeon De Witt appointed his associate James Geddes to carry out this mission of the legislature. In 1808, Geddes was already a mature man of forty-five. Having grown up in the Syracuse area, he began his career as a manufacturer of salt, then served as justice of the peace and later as a member of the Assembly. He would subsequently be elected to Congress and return to the Assembly before assuming high duties on the Erie Canal.

Local judges in those days were frequently called upon to settle property disputes and Geddes, like many of his colleagues, became a skilled surveyor in the process. Geddes might well have spurned the $600 honorarium De Witt offered for this job, but he was already an enthusiast for a project along the lines of Forman's proposals, as we have seen from his earlier discussions on the subject with Gouverneur Morris and Jesse Hawley.

The surveyor general, however, favored the two-canal system as proposed in Gallatin's report, with one canal from the Hudson to Lake Ontario and the second connecting Lake Ontario to Lake Erie. Consequently, De Witt urged Geddes to spend most of his time studying the Ontario route, with just a quick survey of the interior route recommended by Forman. Full analysis of the western lands could be done later, De Witt went on, in the event that the government deemed it necessary.

Geddes spent the summer of 1808 following De Witt's instructions to study once again the well-traveled area between the Mohawk and Lake Ontario. That effort used up all of the $600 appropriation, but Geddes's deep personal interest in the matter persuaded him to push on anyway, in order to survey the lands as far west as the Genesee River. Geddes was particularly interested in exploring the lands just east of Rochester, toward Palmyra (then known by the less attractive name of Mud Creek). This project entailed the outlay of an additional $73 out of his own pocket, for which the legislature subsequently reimbursed him.

This was $73 well invested. Until Geddes explored this area, it had been widely assumed that the territory consisted of ground situated so far above the level of the rivers and lakes that there would be no source of water to supply a canal. Trekking over snowy hills, Geddes came to an area known by the romantic Indian name of the Irondequoit Valley, just to the east of the town of Rochester. There he was both surprised and overjoyed to discover that the course of the Genesee River was well above the heights on either side of the valley and could easily feed a 500-yard embankment about seventy feet high to carry the canal between the hills.

It was a big moment. As Geddes recalled the occasion in a letter to a friend fourteen years later, "I had, to be sure, lively presentiments, that time would bring about all I was planning, that boats would one day pass along on the tops of the fantastic ridges, that posterity would see and enjoy the sublime spectacle. . . . There are those, sir, who can realize my feelings on such an occasion, and can forgive, if I felt disposed to exclaim Eureka, on making this discovery. Boats to pass over these arid plains, and along the very tops of these high ridges, seemed then like idle tales to every one round me."[19] By the date of this letter, Geddes's vision was already a functioning reality.

* * *

While Geddes was tramping his way through the wilderness from the Mohawk to the Genesee, Simeon De Witt back in New York City was investigating the feasibility of extending the canal through the lands lying even farther west, across the 75 miles or so between the Genesee and Lake Erie. Most of the land in this enormous area of 3.3 million acres was owned by a commercial operation called the Holland Land Company, a consortium of six Dutch banks.[20]* Their purchase was so large it covered almost the whole of what is today Genesee County.

The company had originally acquired the properties from Robert Morris in 1797, when the financier was desperately raising cash to pay off his creditors. The plan was to break the single large holding into small lots and sell them off to settlers moving in from the Atlantic seaboard. Selling prices would settle at around $160 for quarters of an acre in the villages and $600 for forty-acre lots in the countryside.[21] This was an ambitious objective in 1800. Only about 17,000 people were living in the area, of whom a mere 1300 were over forty-five and nearly half were children under the age of seventeen.[22]

The Holland Land Company's resident agent in the territory was Joseph Ellicott, who had established the company's operational head-

*General Schuyler and the New York State authorities had tried their best to lure the Holland Land Company into investing in the Western Inland Lock Navigation Company. The story is a long one, but the company never did invest in the Western Company. For full details, see Nathan Miller, *The Enterprise of a Free People*, pp. 26–29.

quarters in the little town of Batavia, about forty miles east of Lake Erie. Born in 1760, Ellicott was more than six feet tall, hardy, and toughened by years of work in the outdoors as a surveyor. He had grown up in a close-knit family with a passion for mathematics. His father, a master clockmaker, had taken the family into the wilderness in the area of Chesapeake Bay, where they had built their home with their own hands. All of this contributed to Ellicott's skills as a surveyor.

Ellicott worked at surveying all his life and never lost his enthusiasm for mathematics, but he also turned out to be a first-class salesman and marketing man, a clever if amateur lawyer, and an astute manipulator of political power to advance the interests of his company. Yet his personal integrity was never questioned. Ellicott's Dutch superior in Philadelphia, Agent General Paul Busti, gave him wide latitude in making decisions and disbursing funds.

Ellicott had every reason to support the idea of a canal either through or close to his properties. The roads through his territory were mostly Indian trails with little capacity for carrying the heavy cargoes of farm output. Improved transportation was the only hope of attracting enough people to populate the vast area. Upon receiving De Witt's inquiry, Ellicott estimated that the Holland Company's property would increase to $1 million in value if a canal was built through the region.[23]

Although Ellicott did not provide De Witt with a detailed survey of the territory, he did supply a description and, even better, a map. He also supplied his condition for permitting the canal to cross his lands and for his company to contribute both money and property to its development: he wanted nothing to do with the Gallatin proposal of two separate canals; he would support only a direct water route from Albany to Buffalo. Finally, he offered his estimates of required water depths, number of locks, and excavation costs along his recommended route.[24]

It was now late 1808, the year the Gallatin report had been published, and New York's Surveyor General Simeon De Witt had all the material he needed to make a presentation to the president of the United States. He promptly remitted his full report to Washington.

Silence.

* * *

Joshua Forman, who had started this ball rolling almost a year earlier, was not to be put off by the president's disregard for De Witt's carefully assembled documentation. In January 1809, Forman traveled down to Washington on his own to obtain an appointment with Jefferson. His congressman made the arrangements, and, after all the suitable introductions had been made, Forman began his discourse. He explained to the president that the New Yorkers—or, in Jefferson's lingo, the Yorkers—had found the construction of a canal cross-country from the Hudson to Lake Erie "practicable beyond their most sanguine expectations." Forman then went on, "in as laconic a manner as I could," but doubtless at great length, to recite the many advantages such a canal would bring to the nation as a whole.[25]

Jefferson's response was crushing. All the noble words about public improvements and spending the surplus "in time of peace [for] rivers, canals, roads" seem to have been forgotten. Jefferson may have been concerned about the deterioration in the federal financial position as a result of the trade embargo he had just imposed as among the first steps leading up to the War of 1812. But he was surely consumed by frustration that the devoted efforts of George Washington and the Patowmack Company in Virginia had been far from a brilliant success, and now—as he had predicted to Washington a quarter of a century earlier—the Yorkers were about to beat them to the punch. "Why sir," he declared to Forman, "here is a canal of a few miles, projected by General Washington, which, if completed, would render this a fine commercial city, which has languished for many years because the small sum of 200,000 dollars necessary to complete it, cannot be obtained from the general government, the state government, or from individuals—and you talk of making a canal of 350 miles through the wilderness—it is little short of madness to think of it at this day."[26]

Forman proudly held his ground. His parting words to the president were that "the state of New-York would never rest until [the canal] was accomplished."

CHAPTER 7

THE EXTRAVAGANT PROPOSAL

The die was cast. Jefferson had put the Yorkers on their own. Joshua Forman's proud riposte to the president made it clear they were ready to accept the challenge.

But not right away. The idea of an artificial waterway from one end of New York state to the other still seemed beyond the reach of reality to most of its citizens. The catcalls and ridicule that greeted Forman's address to the legislature in 1808 represented widely held misgivings. Thomas Jefferson had axes to grind, but, as he would concede later, his sense of incredulity was real. As we have seen, Elkanah Watson himself had confessed the idea struck him as "extravagant as the possibility of a canal to the moon." Even the hard-nosed managers of the Holland Land Company, who had everything to gain from a canal across their territory, were constantly dithering over how far they should commit themselves to contributions of land and money because of nagging doubts that the whole thing would ever come to pass.

But a tougher and more visible barrier than disbelief was about to interrupt progress toward the canal. Americans were suddenly confronted by a grim predicament. Thousands of miles of ocean were insufficient to keep them isolated from the messy world of Europe, the world they thought they had left far behind when they or their forebears sailed to these shores with high hopes for a life of their own making. The Revolution had been created from these hopes, but the separation from Europe's affairs was destined to be less complete than the Americans had anticipated.

Trade was the source of the trouble. In peacetime, trade is such a powerful force it can tatter boundaries and shrivel oceanic distances. In wartime, freedom of the seas becomes one of war's first victims. As the Napoleonic Wars raged throughout Europe, the trade routes across the Atlantic turned into a major battleground. American shipping was soon caught in the middle, treated by both sides as little more than an interloper in the larger affairs of the world. With trade so disrupted, the very purpose of the new canal across New York was in danger.

Blockades of incoming shipping by the British against France and by the French against Britain wreaked havoc just at a time when both sides were desperate for the output of American farms and factories. As a result, these blockades grew increasingly elaborate. At one point, the British barred all ships from going to France unless they stopped first at English ports and paid the necessary duties. The French retaliated by declaring that any ship obeying the British proclamation would be denationalized and therefore liable to capture and confiscation.[1]

The blockades were only part of the motivation for the British navy's kicking and cuffing against the Americans. The more immediate incentive developed from severe problems of morale among its crews. Navy service for the lower ranks meant obnoxious living conditions, low pay, and physically demeaning labor. Men were flogged for the slightest transgressions, food was appalling, quarters were primitive. Many British sailors responded to these conditions by deserting in order to find employment on American vessels. The British navy commanders, facing an increasingly acute shortage of sailors, were equally determined to get their men back on board under their own flag.

The British now started a pattern of constant harassment of American merchant ships. Under the guns of British warships, the Americans had no choice but to allow British naval officers—"down to the most beardless midshipman," as John Quincy Adams would describe it—to come aboard in search of their errant crewmen. Any unlucky deserters were forcibly dragged off and forced back into duty, a practice known as impressment. The British also showed no hesitation in dragging American sailors along to help man their ships. Adams reported that more than six thousand Americans were seized in this manner.[2]

Infuriated Americans demanded retaliation against both the hits to their pocketbooks and the insults to their national pride. President Jefferson, however, was opposed to any steps involving gunfire. With the backing of his old Virginian colleague James Madison, now secretary of state, Jefferson looked for some effective means of response short of opening fire on the British or the French. The result was one of the most bizarre pieces of legislation in more than two centuries of American history: the Embargo Act of December 22, 1807, which mandated an embargo on all ships and vessels in U.S. ports and planning to head out to any foreign ports of places. The act essentially outlawed all exports from the United States to any country in the world, on the theory that the Europeans needed American food more than Americans needed European luxuries. In other words, time had to be in favor of the Americans.

But waiting for time to set matters to rights came with a high cost. Economic activity came to an abrupt halt. New York, as a major port, was especially hard hit. An English traveler passing through the city four months later observed that "every thing presented a melancholy appearance. . . . The few solitary merchants, clerks, porters, and laborers that were to be seen were walking about with their hands in their pockets. The coffee-houses were almost empty; the streets, near the water-side, were almost deserted; the grass had begun to grow upon the wharfs."[3]

Even in the case of the Holland Land Company's holdings in western New York, four hundred miles due west from the Atlantic, sales of lots to settlers dried up to practically nothing. Many of those who had already bought land had done so on the basis of ten-year installment contracts, including two years without interest payments.* With the European markets for American food blocked off, they were no longer able to maintain their monthly remittances.

There were offsets. Smuggling developed rapidly and compensated for some of the official loss of international commerce. As always, necessity is the mother of invention. Up to this moment, traders in Montreal had imported a wide variety of products like flour, pork, whiskey, and potash from the American coastal cities, with most of it coming up from New York City. When the embargo shut off all shipments out of New York harbor to foreign lands—and Canada was a foreign land—the settlers in Holland Land territory lost no time taking advantage of the situation. They promptly opened up new trade routes between western New York State and Montreal across Lake Ontario and up and down the St. Lawrence River. Resident Agent Ellicott was delighted by the improving financial position of the families who had bought land from him, even though there were moments when his conscience bothered him over smuggling as the source of the cash flows.

The loss of textile supplies from Britain led to the birth of America's own textile manufacturing industry, an unexpected outcome nurtured by mercantile capital no longer involved in financing foreign trade. In 1814 alone, 105 new textile mills would open in the United States. In one of the ironies of history, Jefferson's embargo encouraged the industrialization he feared and that his prime opponent, Alexander Hamilton, spent his career in promoting.

Time proved that Jefferson's Embargo Act was untenable. It inflicted economic pain, it made unlawful activity profitable, and it failed to change the situation at sea. The British and French went about their business as though nothing had happened. In March 1809, Congress

*Is this where the American automobile companies got the idea?

repealed the Embargo Act and replaced it with a different but in many ways equally nonsensical piece of legislation, with the awkward name of the Non-Intercourse Act.

This was a kind of 180-degree reversal of the Embargo Act. The Embargo Act outlawed merchandise exports but admitted imports. The Non-Intercourse Act lifted the embargo on exports but outlawed merchandise imports from Britain and France.

* * *

With all this economic turbulence and political uproar, it was early 1810 before New York State could rally enough forces to restart the effort for a waterway across the state and make good on Forman's challenge to Jefferson. When New York finally did spring into action, the provocation came from an unexpected source. The Western Inland Lock Navigation Company was facing another of its endless series of financial crises. The company was spending so much money merely to maintain its existing improvements to Mohawk navigation that nothing was left over from the toll revenue for additional improvements or to pay dividends to the shareholders. As a private enterprise, however, it could not stand still financially forever.

Thomas Eddy, who had been one of the first directors of the company and was now the treasurer as well, had a plan to pull the company out of this bind. Eddy had been trying to persuade his fellow directors to extend their operation westward, with a canal from the headwaters of the Mohawk to the Seneca River—a distance of about seventy-five miles—because he was convinced this opening would increase the flow of toll revenues by several orders of magnitude. His arguments got him nowhere. As he described it some years later, "no importunities . . . could prevail on the company to make advances for further improvements."[4]

In March 1810, inspiration struck. As Eddy recalled the moment, "It occurred to me that possibly the legislature might be induced to appoint commissioners to examine and explore the western parts of the state, for the purpose of ascertaining the practicability of extending canal navigation, and to estimate the expense. . . . I was perfectly convinced that if commissioners should be appointed, they would make a very favourable report."

He immediately turned to his friend State Senator Jonas Platt, who represented the western part of the state and was also the leader in the Senate of Eddy's political party, the Federalists. Eddy could not be sure of what Platt's response was going to be. Platt had many reasons to oppose the canal. He had always taken a dim view of the privately owned Western Company, and Eddy's scheme was frankly to develop public support for his company. Platt had never forgotten that General Philip Schuyler and William Weston, the British engineer involved, were as knowledgeable about waterways as anyone in America at that time, but their forecasts of the company's expenses and tolls "erred more than 200 percent in their estimates."[5]

There was also the record of the Middlesex Canal, at that time the longest canal in North America, a twenty-seven-mile waterway built in the late 1700s to connect Boston to the Merrimack River in the northern reaches of Massachusetts. The Middlesex did a good job of moving heavy material like granite and lumber but was never able to generate enough revenue to stay current on its debts. If Schuyler and Weston could be so wide of the mark on a relatively simple undertaking, and if the Middlesex was such a financial failure, what confidence could people place in anyone who recommended a project as large, as complex, and as novel as the Erie Canal?

To Eddy's surprise, none of this discouraged Platt. He decided that a small group of high-powered commissioners reporting to the legislature would be a shrewd move in overcoming the credibility problem and persuading the skeptics the job could be done, especially if he could provide the group with a strong and highly visible leader. He was also determined to go further than Eddy's proposal of extending navigation just seventy-five miles beyond Rome. Platt told Eddy he could foresee a state-sponsored canal running all the way to Lake Erie and using the rivers only as feeders, although he would leave it to the commission to determine whether to go directly overland to Lake Erie or to make use of Lake Ontario.

Platt took a firm position on another matter of great importance that would shape the entire future of the Erie Canal: "No private corporation was adequate to, or ought to be entrusted with, the power and control over such an important object." New York State itself would do the financing. Finally, he declared to Eddy, the less the Western Company had to do with the whole business, the better the chances of getting somewhere with both the legislature and with public opinion.

As Platt recollected their conversation, he claimed that Eddy, "that prudent and excellent man, seemed startled at the extravagance of my proposal . . . so visionary and gigantic the legislature would not even deem it worthy of consideration." Platt needed hours of persuasion before his friend was willing to accept the practicality of his enlarged proposal. But by the morning Platt had his resolution ready for the legislature.

Platt and Eddy proceeded to identify seven commissioners they would recommend, an artful job of combining, as Platt put it, talent, influence, and wealth. They also suggested that James Geddes serve as an expert to accompany the commission, in particular to show them the route he had found on his own recent expedition west. Geddes would be accompanied by his good friend the surveyor Benjamin Wright, who was sharing living quarters with Joshua Forman in 1808 when the latter launched his frustrated campaign for a canal to Lake Erie.

All but two of Platt and Eddy's recommendations for commissioner were shareholders in the Western Inland Lock Navigation Company. There were four Federalists, the party of the wealthy upper classes and followers of Washington and Hamilton as supporters of the dominance of the federal government over the states. They were Gouverneur Morris, the well-known statesman and the owner of extensive properties in the valley of the Genesee River; Stephen Van Rensselaer, New York's largest landowner (widely known as the "patroon"), highly popular, and one of Watson's companions on his trip west in 1791; William North, a respected veteran of the Revolution and heir to the rich lands of General Friedrich von Steuben's estate; and Thomas Eddy.

There were also three members of the opposition party, the Republicans—not the Republicans of our own time, but the forebears of the Democratic Party, whose ideas reflected Jefferson's view of the states as the true seats of power combined with a vision of the needs and rights of "the common man." These party characterizations would not strictly hold up as the years went by, and De Witt Clinton in particular was on many occasions dependent on Federalist support even more than he could expect votes from his own Republican Party. In 1810, the lines were still clearly drawn.

The other two Republicans were De Witt Clinton's first cousin Surveyor General Simeon De Witt and Peter Porter, a congressman with large business and property interests in western New York.

Although Porter was officially a member of the Republican Party, he was no friend of Clinton's. He was a strong supporter of a waterway to the west in large part because of the economic benefits he could foresee for himself. At the moment when Eddy and Platt were attempting to put the group in place, Porter had just made an eloquent maiden speech to the House of Representatives on using public funds for both roads and canals. His efforts resulted in a bill "for the improvement of the United States by roads and canals [including] opening canals from the Hudson to Lake Ontario and around the Falls of Niagara."[6] Like previous efforts and future efforts to come, and despite the Gallatin report (on which it drew), Porter's proposal died in Congress as sectional biases continued to prevail over a national perspective. New York was still on its own.

* * *

Although all their plans were now in place, Eddy and Platt were not yet ready to go to the state legislature. They were certain their strategy would go nowhere without the unqualified support of De Witt Clinton, but here, again, the outcome was uncertain. Not only did they sit on opposite sides of the aisle, but Clinton had recently referred to Platt sarcastically as "the would-be Governor."[7] More important, Clinton had never given much thought to the matter of a canal to the west, although, at Columbia College he had studied under a Scotch mathematics professor who gave his students many lectures on the history and principles of canal building.[8]

With all their arguments carefully organized when they met with Clinton, Eddy and Platt also brought along Geddes's 1808 report with its unqualified conclusion that a continuous canal across the state was feasible. Platt put it well when he observed that "Mr. Clinton" now grasped the canal project "as an object of the highest public utility, and worthy of this noblest ambition."[9]

When Clinton made the canal the central focus of his long and distinguished political career, he created such a powerful wave of popular support for the project that in time even his bitterest political enemies no longer had the courage to oppose him on this issue. Even then, his adversaries had no intention of abandoning the attack on him. Determined to assault him no matter what, they would ultimately turn the canal controversy on its head and claim that they, rather than Clinton, were the true champions of this noble cause.

* * *

De Witt Clinton's political achievements alone would mark him as an outstanding American. In New York, he had been secretary to the governor, his uncle George, who, after distinguished military service in the Revolution, held the post from 1777 to 1795. Although De Witt had represented New York as a U.S. senator for part of one term, from 1802 to 1803, he had led the forces there for the enactment of the Twelfth Amendment, which mandated the separate election of president and vice president. But Clinton was not happy in Washington, which was still more swamp than city and where his only social life was with government people. He missed the lively action and the intellectual companionship he so much enjoyed in New York City. He sensed that developments there were likely to matter more to the nation than decisions taken in Washington.

The perfect opportunity to pull out and resign from the Senate came along in 1803, when—probably to no one's surprise—Governor George Clinton appointed him mayor of New York City, an office that would advance De Witt's political ambitions while helping to support his own power against a rising threat from Tammany Hall, a newly established grassroots organization eager to gain control of the electoral process in New York. De Witt Clinton's only other foray into the world of Washington came in 1812, when he ran for the presidency against James Madison on a peace ticket.

He served a total of ten one-year terms as mayor—1803–1807, 1808–1810, and 1811–1815—which included periods when he was also lieutenant governor and simultaneously represented the city in the state Senate. He also served twice as governor of New York State, first from 1817 to 1823 and then from 1825 until 1828, when he died in office. The period from 1811 onward was critical in the launching of the Erie Canal.

As mayor, Clinton had wide powers and many opportunities to dispense patronage in the form of such jobs as licensed marshals, porters, carriers, cartmen, cryers, scavengers, as well as licensing tavern keepers and anyone selling liquor at retail. He could remove any of these individuals at his pleasure.[10] He also served as presiding judge of the Court of Common Pleas, a responsibility he took seriously and exercised with notable harshness on many occasions.

Although Clinton was born and raised at Little Britain, near Newburgh in the Hudson River valley, he had been enthralled with New York City, seventy-five miles to the south, from his earliest boyhood. New York had a special interest, because so much of it was new. In September 1776, as Washington's forces were beating a hasty retreat before the advancing British, a fire broke out in the middle of the night, destroying nearly half the city west of Broadway; six hundred of New York's thousand homes burned to ashes. In 1803, that whole rebuilt area was as modern as any community in the whole country and already on its way to becoming the key national melting pot.

Clinton especially enjoyed the motley crowds of people who poured into the budding metropolis. Favorite jaunts in those days, when time allowed, included a trip to the charming hills and river views in Harlem Heights—another scene of colonial defeat by the British invaders.

Life in this lively and dynamic community was so tempting for a young man that Clinton had passed up Princeton, the favorite university of his relatives, to stay in New York. In 1784, at the age of fifteen, he would be at the head of the line to sign up for the first class to pass through Columbia College—the class of '86—after its establishment on the dilapidated premises of the old King's College.

By the time Clinton became mayor, New York, with its magnificent harbor, had overtaken Philadelphia as the nation's largest exporter and importer. He was determined to make New York a great city, a safe and attractive place for its 90,000 inhabitants to live, and the leading source of the nation's economic power. He spent his greatest efforts to establish free schools for the poor. In addition to his duties as judge, he chased after almost every fire alarm to supervise the performance of the volunteer fire department, and participated in the creation of the city's first fire insurance company. As his mayoral responsibilities placed him at the head of the police force, he personally put down several dangerous riots in the streets. When yellow fever struck in 1809, and many city officials and common citizens retreated a mile or so northward to the village of Greenwich, Mayor Clinton remained in the city conducting business as usual. He obtained a charter for the Society for the Manumission of Slaves in New York State, put through legislation removing all political and social disabilities of Roman Catholics in New York City, and promoted the establishment of a free hospital for the insane on an estate known as

Bellevue, the forerunner of Bellevue Hospital. He gave his full backing to Robert Fulton's projects for a steamboat company to ply the Hudson River all the way up to Albany.

All of these efforts attracted increasing numbers of people to the city. In 1810, seven years after Clinton became mayor, Manhattan's population had increased by more than 40 percent over the last decade, to 125,000; it was now the largest city in the country. But the population of the rest of New York State was also exploding, having grown from 589,000 in 1800 to 959,000 by 1810.[11] Another 400,000 people would be added in the next decade as well, even as New York City's head count was rising by only 37,000.*

There was a lot more to Clinton than his powerful political drive. He was a serious scholar of history and student of nature, especially in the field of ornithology, which held a long fascination for him. He had the commanding physical presence of a natural leader—tall and broad, with a notably high forehead, clear eyes, and a set expression around the mouth. He was never someone to be trifled with, as his many enemies would learn to their sorrow.

His first wife was Maria Franklin, a wealthy and beautiful woman with whom Clinton had ten children, only one of whom, the oldest and his favorite, died in childhood. They entertained frequently at their home at 52 Broadway and associated with all the wealthy and powerful people of their era. Meanwhile, Clinton was a devoted father, spent much time with his children, did all the family bookkeeping, and even found time to do the family marketing.

How could he manage it all—marriage, children, politics, and his broad interests in the arts and sciences? The Clintons had unlimited service available from grooms, valets, and maids at home, as well as secretaries and other subordinates when he went off to work. He evidently

*At the time of the Revolution, Virginia, with almost 700,000 people, was by far the most populous state, nearly double Pennsylvania, which was in second place, and far ahead of New York.

took all of this for granted, for none of the massive accumulation of historical and biographical material about him makes any mention of these individuals.

The full flavor of this amazing man, his mind, his erudition, and his eloquence, comes through in this excerpt from a lengthy address (sixty printed pages!) he delivered in 1811 before the New-York Historical Society on the Iroquois, describing in great detail their history and the high level of their civilization:

> Indeed, when we consider, that the discovery and settlement of America, have exterminated millions of the red men, and entailed upon the sable inhabitants of Africa, endless and destructive wars, captivity, slavery and death, we have reason to shudder at the gloomy perspective, and to apprehend that, in the retributive justice of the Almighty, "there may be some hidden thunder in the stores of Heaven, red with uncommon wrath,"* some portentous cloud, pregnant with the elements of destruction, ready to burst upon European America, and to entail upon us those calamities which we have so wantonly and wickedly inflicted upon others. . . .
>
> And, perhaps, in the decrepitude of our empire, some transcendent genius, whose powers of mind shall only be bounded by that impenetrable circle which prescribes the limits of human nature, may rally the barbarous nations of Asia under the standard of a mighty empire. . . . And if Asia shall then revenge upon our posterity the injuries we have inflicted on her sons, a new, a long, and a gloomy night of gothic darkness will set in upon mankind. And when, after the efflux of ages, the returning effulgence of intellectual light shall again gladden the nations, then the wide-spread ruins of our cloud-capp'd towers, of our solemn temples, and of our magnificent cities, will, like the

*A quotation from Joseph Addison's 1713 tragedy *Cato*.

works of which we have treated, become the subject of curious research and elaborate investigation.[12]

* * *

By late June 1810, less than three months after the New York State legislature gave its blessing to Platt and Eddy's proposal, the commissioners had organized their respective responsibilities and were on their way west. Gouverneur Morris, as the oldest and probably most famous member of the group, was given the title of senior member. Eddy was appointed secretary and treasurer, titles that turned out to include master of entertainments, supplier of "segars" and wine both red and white, as well as transporter of a library that filled an entire trunk.

All members other than Van Rensselaer and Morris were to go by boat up the Mohawk and then continue west by water as much as possible, until they reached Geneva on Lake Seneca. They would sell their boats at that point and proceed by carriage over the remaining hundred miles to the Niagara River and Lake Erie. Up near the headwaters of the Mohawk, James Geddes and Benjamin Wright would join them as surveyors. Peter Porter planned to meet the group upon their arrival in Buffalo. As background and research material, the commissioners brought along Geddes's survey of 1808, Ellicott's letter and map, and Jesse Hawley's essays. Presumably these plans envisaged exploring at least in part the alternative of a short canal heading only from Albany to Lake Ontario and then another even shorter canal along the Niagara River into Lake Erie.

Meanwhile, Van Rensselaer and Morris would cross the whole state by land, right out to Lake Erie. Although Morris was now fifty-eight, the prospects of such a strenuous excursion held no concerns for him. On the contrary: he would bring along his new young bride and his French cook, in addition to a painter to record the scenery they were going to pass through.

Morris was a remarkable man, the full personification of great vitality. As a young teenager, he had suffered ruinous burns from a boiling kettle that spilled over his right arm and side. Then he lost a leg in a carriage accident when he was twenty-eight.* The peg leg that replaced his own never slowed him down, nor did he use it as an excuse for anything. He was as active physically (in the fullest sense of the word) as any man, and had frequently traveled cross-country over the roughest terrain. With his round face, pointy nose, and slightly receding chin, he continued to follow the outmoded fashion of powdering his hair, an affectation despised by Jefferson—who had his own outmoded affectation in referring to his countrymen with the French Revolution's pet word of *citoyens*.[13]

New York City was Morris's great love. In 1811, he had participated with Mayor De Witt Clinton in designing a grid covering all of Manhattan Island above Houston Street, 12 avenues running north-south and 166 cross streets, plus the diagonal of Broadway—an old Indian trail. Except for the expanse of Central Park, visitors to New York today would have little trouble finding their way around Manhattan with the Morris-Clinton ground plan.[14]

* * *

The commission of 1810 reflected a profound shift in both the proposed structure of the canal and, indeed, in the very perception of the canal. It was not just that private enterprise was out and government ownership, sponsorship, and management were in. The key development was that now the Erie Canal had ceased to be just a topic of conversation and occasional exploration.

Platt's strategy had worked. With names like De Witt Clinton, Gouverneur Morris, Simeon De Witt, and Stephen Van Rensselaer taking charge of matters, people were now willing to place confidence in the notion of a canal over mountains and valleys over a route more than 300

*John Jay, upon hearing of the accident, wrote to a friend that he wished Morris "had lost something else." See Richard Brookhiser, *Gentleman Revolutionary*, p. 61.

miles long. There were to be no more expressions of surprise and ridicule, and work could finally begin on the real thing. The War of 1812 would interrupt progress and lead to a few dark days of discouragement near its end, but from here forward nothing would reverse these decisions or draw them into question ever again.

CHAPTER 8

THE EXPEDITION

De Witt Clinton and Thomas Eddy traveled together from New York to Albany in early July 1810, where they would join the rest of the newly appointed commissioners. Although they could have taken the usual sloop up the Hudson River to Albany, there was a more tempting alternative: Robert Fulton's steamboat, *The North River Steamboat* (later known as the *Clermont*), which had been running the New York–Albany route for nearly three years. They signed up for the steamboat, looking forward to the historic breakthrough of traveling on a boat that could go upstream as well as downstream without sails, oars, or poles.

The North River Steamboat was scheduled to leave for Albany at noon on June 30, but Clinton nearly missed the boat while waiting for a new servant to whom he had paid a month's wages in advance. The man never showed up. Clinton must have started off on this adventure in a high state of aggravation. All he has to say in his exceptionally verbose journal about one of history's early steamboat rides was that the weather was warm and the boat was crowded.

The voyage must have been amazing. Fulton's creation was a paddle wheeler, 149 feet long, almost 18 feet broad, and 182 tons in weight.

These boats were no objects of beauty. An English traveler as late as 1830 described them as "the least picturesque [of vessels]. . . . Their smoking chimneys[, and] their ungraceful and worse than dromedary projections, give the idea of a floating foundry."[1] Nevertheless, belching copious quantities of flame and steam, these floating foundries swished their way along at four to five miles an hour, much faster than poling or rowing and more certain and direct than a sloop in their passage.

The interior height of *The North River Steamboat* was 6½ feet, "sufficient for a man with a hat on," as Fulton described it (and most hats were stovepipes).[2] The fare was $1 for every twenty miles and $7 for the full 150-mile trip to Albany—about what the 152-mile trip from Utica to Rochester on the Erie Canal would cost in 1825. Anticipating the Mississippi paddle wheelers, to say nothing of the great ocean liners of the future, Fulton had designed his vessel to be, in his words, "a floating palace, gay with ornamental paintings, gilding, and polished woods." There were provisions for fifty-four berths divided among three cabins, as well as cooking facilities and a bar. Dinner was served at two o'clock, followed by evening tea with meats at eight o'clock, each meal costing 50¢. An awning provided cover from rain and sunshine, so that the passengers could "dine in fine weather and the place is so spacious it will be charming. . . . All dirt [will be] out of sight."

Berths were assigned in order of arrival. Fulton set up rigid rules of decorum, with fines for breaching them, such as $1.50 for the first hour anyone lay down with his shoes or boots on, plus 50¢ for each additional half hour "they may offend against this rule." The fines financed wine for mealtimes. Smoking was forbidden belowdecks and women were forbidden to smoke anywhere.* Card games were to end at 10:00 p.m. so as not to disturb sleepers.[3]

Although Clinton's journal tells us nothing of substance about his voyage up the Hudson, it would be interesting to know what went

*Women were smoking in public places in 1810!

through his mind as *The North River Steamboat* chugged along toward Albany. Did he think about the glorious scenery he passed by, such as the steep palisades on the New Jersey side just north of New York City? Did he recall the British and Hessian troops who had scaled those heights in pursuit of Washington's army in late 1776—and failed in their mission? Or, about seventy-five miles farther on, did he take note of Newburgh on the western bank of the river, just eight miles from Little Britain, where he was born forty-one years earlier and where the purplish haze of the Catskill Mountains first becomes visible to the traveler heading north? Did he recall that George Washington was at Newburgh in 1783 when word arrived that Britain had signed the Treaty of Paris and recognized the independence of the United States of America? Did he admire the point and the steep cliff sticking out into the river as the steamboat passed West Point, the home of the young republic's military academy? Did he think about the remarkable character of the Hudson itself, which is not really a river but a great estuary of the Atlantic Ocean, reaching nearly two hundred miles into the interior of New York State with ocean tides felt even at the most northern reaches of the river? Did he ask amuse himself by considering that the great project inspiring this expedition would in all likelihood bypass the entire length of the Mohawk River but that nobody would dream of ever bypassing the mighty Hudson?

Whatever his thoughts, Clinton had to put up with about forty hours of the crowd and the heat he complained about, as well as Fulton's elaborate rules and regulations, until *The North River Steamboat* arrived in Albany shortly before daylight on Monday, July 2. After finding accommodations at an inn, he and Eddy joined the other commissioners at the office of Clinton's cousin, Surveyor General Simeon De Witt, to settle on the plans for the trip up the Mohawk and then across the western half of the state. The group spent the afternoon searching for the equipment they would need in the days ahead—a mattress, a pillow, and a blanket for each. Clinton later complained that they neglected to provide themselves with "marquees and camp-stools, the want of which we sensibly experienced."[4]*

The trip they were about to commence would turn out to be both more and less than they anticipated. They would experience frustration and excitement, as well as fun, hilarity, boredom, and bugs. In many ways, the trip sounds like an adventuresome jaunt by a bunch of men liberated from the constraints of office work for the joys and the freedom of the great outdoors. On the serious side, the commissioners would arrive at a sequence of important decisions that would set the foundations for their future work. Controversy, however, would persist.

* * *

The group—William North, Simeon De Witt, and James Geddes, in addition to Clinton and Eddy—departed by carriage for Schenectady at five the following morning, expecting to find their bateaux waiting there to transport them up the Mohawk. Then they discovered that Eddy had failed to order the boats, so the group had to spend the Fourth of July in Schenectady. Once again, Clinton was in ill temper. Always impatient, he scorned Schenectady and could not wait to leave it: "The true reason for our anxiety was the dullness of the place. Imagine yourself in a large country village, without any particular acquaintances, and destitute of books, and you can imagine our situation." The commissioners passed "the gloomy interval" in Schenectady by watching the local holiday pageantry but found nothing to please them in the show.[5]

Poor Schenectady—S'coun-ho-ha! The Dutch had settled the town, the first of them exclaiming on his arrival in 1642 that this was "the finest land the eyes of man had ever rested upon."[6] Schenectady suffered a horrible massacre in 1690, when Indian raiders murdered or took prisoner every single resident of about sixty houses and then set each of the

*A marquee is in all likelihood a tent, although it may have been a simple stand to provide shade.

homes ablaze. The entire community would succumb to fire once again in 1819, when the Erie Canal was already two years into construction.

To their happy surprise, the commissioners found two bateaux were ready and waiting for them by four o'clock in the afternoon, with one for the commissioners and one for their baggage. The passenger boat was furnished with an awning, curtains, and enough seats for the four commissioners (Van Rensselaer and Morris would meet up with them farther west) and the captain. The boat also carried a small crew to handle the sail or to do the poling when conditions made sailing impossible. Three servants and over a ton of baggage were lodged on the other boat.

This was not exactly luxury travel. The men could sightsee and fish from their bateaux, but they could not cook, eat, or sleep aboard. The records are silent, but they doubtless had to go ashore to take care of other human needs as well. Yet a trip by water was certain to be more comfortable than the stagecoach, with its open sides, leather screens let down in stormy weather even in the oppressive heat of midsummer, and uncertain springs as the only cushion against the violent bumps of the roadway; bruises to passengers were not unusual.[7]

The group was enthusiastic about the arrangements. In honor of their leaders, they christened the passenger boat *Eddy* and the other boat *Morris*. As Clinton described their merry departure from Schenectady, "A crowd of people attended us at our embarkation, who gave us three parting cheers. The wind was fair, and with our handsome awning, flag flying, and large sail . . . we made no disreputable appeal."

Adventures and mishaps assailed them almost immediately. They discovered the mast was too high and the ballast too light to hold the boat steady in the face of heavy and sudden gusts of wind. The crew soon had to pull down the sail and the mast and, at great effort, start poling the boat upstream. Then, as evening approached, they discovered they had failed to bring everything with them from Schenectady. Consequently, the first day ended without much progress: they spent the night of July 4 at a small tavern just three miles up the Mohawk from Schenectady.

* * *

Clinton's handwritten journal—all 170 pages of it as we have it today in print—narrates this expedition in a detailed and colorful manner. Quite aside from the careful account of the business of the trip and a precise record of mileage covered from each point to the next, he reports, among other things, on the ancient Indian monuments they found and the meaning of these works for Indian history and culture; on the geology of the country they passed through; on the birdlife and the fish; on the history, features, and inhabitants of nearly every town; and on the taverns, on the meals, on the amounts farmers have paid for the lands and on the amounts they receive for their crops. He lists twenty-three kinds of birds they spotted along the Mohawk and eleven kinds of delicious fish that abound in the clear and uncontaminated waters of Lake Erie.[8] The habitat and behavior of the eels of Niagara Falls interest him more than the gigantic flow of water, foam, and spray. In a typical observation, he describes Chippeway, near Niagara Falls, as "a mean village of twenty houses . . . [but] the most opulent man does not pay more than three dollars a year in taxes. . . . The race of a mill-dam here conceals a boiling spring, which will boil a tea-kettle."

The rivers and the produce of the towns supplied some delicious meals. Just above Utica, the commissioners bought a basket of eggs at a shilling a dozen, excellent butter at fifteen cents a pound, plus nine fish speared from the river and weighing around a pound each, and all this in addition to a small nocturnal heron they shot and a snapping turtle they speared.* But the elements prevented them from full enjoyment of this tempting repast, as a monster thunderstorm suddenly assailed them,

*The reference to both shillings and cents in one transaction was not unusual in those days. Before 1788, each of the colonies, and then the states, had its own currency, denominated in English pounds, shillings, and pence but not 1:1 with British pounds. The customary exchange rate between the U.S. dollar, launched in the early 1790s, and the old New York State shilling was 1:8. Hence, the shilling, or twelve pence, the commissioners paid for the basket of eggs was equal to 12.5¢ in U.S. currency. (I am grateful to Professor Richard Sylla of the Stern School at New York University for this explanation.)

forcing them to moor the *Eddy* under a bank, where they huddled over a cold dinner and sipped hot wine to sustain their spirits.

Clinton provides a veritable handbook of accommodations from one end of the state to the other. There were occasional luscious meals at the inns. At one breakfast, they were offered three kinds of bread, three kinds of fish, plus "fried pork, ham, boiled pork and Bologna sausages, old and new cheese, wood-duck, teal and dipper."[9] Some places were less appealing than others. In one small town in the salt country around Syracuse, after a long and tiring day, the party put up in a house crowded with drunks, including the landlord, his wife, and his son. None of the sleeping arrangements was inviting when the companionship of the bunks included a collection of dirty and obnoxious occupants. Clinton and Eddy finally settled down in a room where they were promised quiet and no risk of bugs. But soon "a thousand villainous smells" like boiled pork assailed their noses, the noisy crowd of drunks filled the social rooms, rats were scratching their way through the walls, dogs crawled in and out from under the beds, and even bats were among the occupants of the inn. To complete the evening's discomforts, Clinton reported that they had been "assailed by an army of bed-bugs, aided by a body of light infantry in the shape of fleas, and a regiment of mosquito cavalry."[10]

Clinton fled outdoors, where the moon was full, with a "blaze of unclouded majesty." As he gazed on the wide view of the hills and the flocks of white geese sporting on the river, he noted an Indian hut and fire on the opposite bank, with the occupants preparing for the day's work of hunting and fishing. He listened to "the bellowing of thousands of frogs in the waters, and the roaring bloodhounds, in pursuit of deer and foxes. . . . My mind became tranquilized, and I availed myself of a vacant mattress in the tent and enjoyed a comfortable sleep of two hours."[11] After departing this town, the commissioners dubbed it Bug Bay, a name that, Clinton predicted, it would retain for a long time to come.

There were also long stretches of boredom. At one point, Eddy—now dubbed Commodore by his colleagues—discovered a towel which he had inadvertently slipped into his pocket at the inn where they had slept the night before. He was instructed to walk back and return it to its rightful place. Although hardly an incident of any importance, the slow progress, the monotonous scenery in that stretch, and the merciless heat made almost any event a welcome interruption to the tediousness and slow progress.

There were thrills of a special kind. At a clear spring a few miles up the Mohawk from Schenectady, the country was so romantically wild, so lacking in any sign of human habitation, that the group recalled the time when all of New York was a land of "roving barbarians and savage beasts." At Little Falls, where the great gorge between the mountains would make the entire Erie Canal possible, they were in awe of the massive rocks of solid granite that created the great gorge through the mountains. "You see them piled on each other, like Ossa on Pelion," Clinton wrote, evoking the classical myth of two huge mountains, one set on top of the other.[12] Many of these rocks were thirty or forty feet thick, and worn over time by the violence of the waters rushing over them or driving them from one position to another into every shape imaginable.

The commissioners were impressed with many of the towns where they stayed overnight, and especially with Utica, which they reached on July 10, six days and eighty-six miles from "the gloomy interval" at Schenectady. There were three hundred houses and 1650 inhabitants in Utica, a town with churches for Presbyterians, Episcopalians, Welsh Presbyterians, and Welsh Baptists. In addition to six taverns, fifteen stores, two breweries, and three printing offices, Utica also had a bank—a branch of the Manhattan Company masterminded by Aaron Burr—a post office, and a state supreme court as well as the county court.[13] Most impressive to Clinton, the townspeople enjoyed a choice of two newspapers.

Clinton takes note of the elegance of some of the houses in Utica as well as the numerous stores and the variety of merchandise they displayed. He was impressed with the bridge over the Mohawk and a turnpike of two miles through the town. Nearby, at a large cheese factory, thirty-six cows produced enough milk for four hundred cheeses a year plus milk to support a great number of hogs. There was also a substantial manufacturing establishment for spinning cotton, with 384 spindles on six frames. Shares in the cotton factories were selling 40 percent over their original offering price, but Clinton noted that the forty young girls employed there had "an unhealthy appearance." As a result of all these facilities, real estate appeared to be priced very high indeed in Utica in 1810: lots, corresponding to double lots in New York City, were selling from $400 to $800. Ten years earlier, a judge named Cooper had bought fifteen acres for $1500; they were now worth more than ten times that price.

Rome, on the other hand, was much less impressive than Utica, even with its excellent position between the headwaters of the Mohawk and the rising of Wood Creek. With seventy houses, it had a post office, a courthouse, and only one church but four lawyers. The town had little appeal for Clinton: "Rome being on a perfect level, we naturally ask from what has it derived its name? Where are its seven hills? Has it been named out of compliment to Lynch [the largest property owner], who is a Catholic?"[14] Rome was in fact originally known as Lynchville. Transforming Lynchville to Rome was minor compared with replacing Mud Creek with Palmyra. Rome did provide one exciting event, when they shot a bald eagle with a wingspan of eight feet and formidable talons.[15]

The group was impressed with the huge salt deposits spread out over a large area about halfway between Albany and Buffalo and centered around what would one day be the city of Syracuse. These resources had provided much of the wealth of the Iroquois and were still an important feature of the local economy. The salt deposits would be even more important after the opening of the Erie Canal, which greatly facilitated the transportation of large quantities of salt both west to Lake Erie and east to the growing industrial economy of New York City.

The commissioners reached their final destination and turnaround point at Buffalo and Lake Erie on August 4, having traveled the 363 miles from Albany in thirty-two days. Here they put up at a tavern, where they were "indifferently accommodated in every respect."[16] Buffalo contained only thirty to forty houses, a courthouse, a few stores and taverns, and a post office, but appeared to be busy as a base for sightseeing. Just about everyone traveling westward toward Ohio and beyond came through Buffalo, and few could resist the opportunity to visit the wonders of Niagara Falls just a few miles away.

On Lake Erie, the commissioners visited the only United States naval vessel on the lake, the *Adams*, a brig of 150 tons and four guns. This ship could travel to Fort Dearborn at the southwestern corner of Lake Michigan and back in two months. The British had two gunboats on Lake Erie, one with sixteen guns and the other with twelve, as well as a fort southwest of Black Rock, a town just north of Buffalo. During the War of 1812, the waters of the lake would become a major battleground between the Americans determined to invade Canada and the British equally determined to prevent them from doing so.

The commander of the *Adams* informed Clinton that vessels drawing up to seven feet of water could continue on from the western end of Lake Erie to "Chaquagy [Chicago] and then up a creek of that name to the Illinois River . . . and so down to the Mississippi."[17] With an uninterrupted canal from New York City to Lake Erie, Gouverneur Morris's dream of 1800 could now come true: the way would be open for travel by water from London all the way to the Mississippi River.

* * *

The seven hundred miles across New York State and back took the commissioners fifty-three days. Now it was time for them to get down to business and settle on the contents of the report they would submit to the legislature. They faced a daunting task. In view of the magnitude of the undertaking and its prospective impact on New York State and, indeed, the entire nation, their report would have to be authoritative, complete, and compelling. They would have to provide specific recommendations, with full supporting arguments, for the enormity of the job ahead, the technology to be employed, the route of the canal between the Hudson and the Great Lakes, and the methods of financing the record-setting outlays that would be necessary. Above all, the report should articulate and underscore the grand theme underlying the whole concept: the necessity of binding the new western communities to the original thirteen states strung out along the Atlantic. As a result, it would take six months, until March 1811, before the report of the commissioners was ready for presentation to the legislature.

During the deliberations of the commissioners, Gouverneur Morris turned out to be the most stubborn obstacle to progress. The group looked to Clinton as their leader, but they had named Morris as the senior member of the commission because he was the oldest and in all likelihood the best known among them. This role was largely ceremonial while the expedition was under way, but as senior commissioner, Morris was responsible for drafting the report of the commission's findings and decisions. He had, after all, drafted not only the constitution of the State of New York, but the Constitution of the United States of America; "We, the people" launch his words, which come down to us today.

Now Morris had every intention of putting his handprint on the report in big, bold letters.

In 1803, when Morris had described to Simeon De Witt his idea of "tapping Lake Erie . . . and leading its waters in an artificial river, directly across the country to the Hudson River," he meant precisely what he said. At an early commission meeting at Rome, Morris urged breaking down the high grounds along the eastern banks of Lake Erie and letting the lake's waters follow the level of the country down as far as Utica, providing uninterrupted navigation for that entire sector of the route without dependence on any supply of water other than the lake. For this purpose, and to avoid the complexity of building locks, Morris proposed the construction a single inclined plane, all of one piece, starting at the eastern shore of Lake Erie and carrying the Erie waters on a downward tilt as far as Utica. The drop to sea level beyond that point is so steep that locks would be unavoidable the rest of the way to Albany.

Imagine a huge downward sloping trough, about five feet deep. The trough would be filled by water pouring in at its western end from the shores of Lake Erie and emptying out into the Mohawk River some two hundred miles to the east. At the average downward slope of six inches per mile specified by Morris, the water would flow steadily eastward but so gently that towing a boat in the direction of Lake Erie, up the inclined plane against the current, would still be feasible.

Nature was not so cooperative. The surveyors could confirm Morris's calculations of the drop in the level of the land from the high point at Lake Erie down to Utica, but they had to emphasize that Morris's six inches to the mile was an *average*. The levels vary wildly from valley to flatland and up and down again. For about forty-five miles east of Rochester, for example, the land drops off steeply at a pitch of 2.8 *feet* a mile, almost six times the slope of the inclined plane. The embankments to support the inclined plane would have to be more than 100 feet high—the equivalent of at least a ten-story building in modern times—at many points over the hundred miles between Rochester and the town of Brewerton on Oneida Lake. To add insult to injury, every single creek, river, and lake on the route would have to be bridged.

Despite all these difficulties, Morris was convinced beyond argument of the superiority of this structure, undeterred by Simeon De Witt's view of it in 1803 as a "romantic thing."[18] When De Witt had reminded Morris that "the intermediate hills and valleys [were] insuperable obstacles," Morris replied "*labor improbus omnia vincit* [essentially, the human mind devoted to improvisation could achieve anything], and the object would justify the labour and expense, whatever that might be."[19] Years later, Benjamin Wright would recall, "I feel very confident he [Morris] had no local knowledge of the peculiar formulation of that part of the state."[20] At the time, Morris had his way. Thomas Eddy reported that the commissioners, "believing that [Morris] knew much more than he really did, and distrusting, perhaps too scrupulously, their own judgment, signed, and therefore sanctioned, [Morris's version of the] Report."[21]

Throughout the discussions, Clinton kept urging unanimity even though he was opposed to Morris's brainchild. He was concerned that the entire enterprise would come to nothing if differing recommendations from the commission encouraged new excuses for argument in an already fractious legislature. With his unshakable confidence in the workings of the democratic process, Clinton argued that common sense would prevail once the whole matter were set before the public.

The westerner, Peter Porter, was the most stubborn in opposing the report, and not only because of the inclined plane proposal. In the deliberations of the commissioners, Porter joined the others in their enthusiasm for an east-west route by water, but he did want to make money out of the project if at all possible. Porter's booming business in Black Rock held a monopoly on trading privileges along the road, or portage, carrying freight around Niagara Falls. As a north-south canal coming down from Lake Ontario right there would be a bonanza for him, he consistently favored the route across Lake Ontario—a route that Elkanah Watson had blessed during his trip west in 1792—instead of the overland route all the other commissioners supported.[22]

A graduate of Yale, Porter had been born in Connecticut in 1773 but moved to Canandaigua in western New York to practice law in 1793. He soon became a strong enthusiast and spokesman for the area. He had settled in Buffalo only the year before the expedition of 1810. Porter was the only participant who would unabashedly pursue his own self-interests throughout the whole process, from beginning right up to the end in 1825. Clinton's observation about him at this moment has a strong sarcastic flavor: "It cannot be supposed that this gentleman was governed by selfish motives on account of his interest . . . but it is proper to say, that his conduct throughout was marked by singular inconsistency."[23]

Despite the underlying validity of this accusation, Porter had been fighting in Congress, where he was serving the first of two terms, for financing of a waterway to the west. In that same year of 1810, he had warned his fellow members of the House of Representatives in Washington, D.C., that the western settlers were in desperate need of a cheap means of transporting their excess production to an Atlantic port, and that the absence of good transportation to the markets was already hampering the settlers and holding back the growth of the entire area.[24]

* * *

Although there were moments when Morris's fantasy prompted one or another of the commissioners to threaten to withhold their signatures from the report, Clinton succeeded in keeping the group in line. After the fact, Clinton did confess to second thoughts about his commitment to unanimity. Writing in 1821, he admitted that the board could have avoided a great deal of the ridicule provoked by its report if its recommendations had been more practical. But Clinton felt they had to defer to Morris, whom he characterized as "a man of elevated genius, but too much under the influence of a sublimated imagination." Clinton's choice of the term "sublimated imagination" may sound odd to modern ears, but "sublimated" does mean diverted from an immediate goal to some higher use. In context, it is clear Clinton means an imagination tending toward the sublime but impractical. In any case, the commission was stuck with the glowing hue of Morris's "sublime" idea. When the members assembled to review Morris's draft, as Clinton tells it, they went out of their way to avoid hurting Morris's feelings, especially as they hoped his proposal would be seen as "hypothetical from its very nature, and a mere gratuitous suggestion."[25]

On everything else, the commissioners quickly reached agreement.

The decision stood—with even Porter's acceptance—that the canal would run over land all the way from the Hudson to Lake Erie. Any other route would defeat the primary purpose of the canal, which was to join the west to the east in one seamless community, as Washington himself had envisioned many years earlier. The Lake Ontario route, which Porter had favored, exposed the Americans to the risk of losing the trade of the westerners to Canada, as much of the traffic moving eastward via Lake Ontario would continue on to the Atlantic by way of the St. Lawrence River and Montreal instead of through New York. This unfortunate pattern was already established in the western part of the state, thanks to the federal government's embargoes and non-intercourses.

With remarkable foresight, the commissioners had no doubt that a canal along the inland route they favored would soon stimulate a rapid rate of economic development along its shores. By assuring a large volume of traffic within the territory of the canal as well as traffic moving from the Hudson all the way to Lake Erie and vice versa, the interior route would fully justify the investment involved.

As any financing by way of Washington was clearly out of the question, the commissioners were emphatic on public financing, ownership, and control of the canal by New York State. They considered this matter to be paramount. Here, even Porter chimed in enthusiastically, delighted at the prospect of having the government pay for providing access by water to his properties and forwarding business in the west. None of the commissioners wanted to repeat the sad experience of the Western Inland Lock Navigation Company, with its constant need for funds. Later, Jesse Hawley would put it that Clinton's determination to make the canal a state undertaking took great courage when others were insisting that "it would require the revenue of all of the kingdoms of the earth, and the population of China, to accomplish it."[26]

A great national interest appeared to be at stake. Although the commissioners considered the risk that private financing might turn the canal into a raw opportunity for speculators, they were more concerned about the difficulty of raising money in a capital-short country like the young United States. Their report sounds remarkably contemporary in their recognition of the significance of the cost of capital: "Few of our fellow citizens have more money than they want, and of the many who want, few find facility in obtaining it. . . . Among many other objections, there is one insuperable: That it would defeat the contemplated cheapness of transportation. . . . Such large expenditures can be made more economically under public authority than by the care and vigilance of any company."[27]

These unqualified assertions favoring public expenditure were among the least controversial features of the commissioners' report. The notion that government spending might tread on private interest was still many years in the future. At that moment, anything looked better than the Western Company.

Through it all ran a hope that the federal government might still participate, because New York would not be the only state to benefit from the canal. Despite the failure of Forman's mission to Jefferson the year before, Gallatin's words from 1805 still rang in the air. Gallatin's powerful report and keen analytical approach had stirred widespread anticipation of the federal government's role in financing and planning a modernized national transportation system. As Gouverneur Morris reflected this state of anticipation in the commission report, "The wisdom, as well as the justice of the national legislature, will, no doubt, lead to the exercise on their part of prudent munificence."[28]

Finally, the commissioners insisted that the canal must be an artificial waterway over its entire distance, no matter which design would finally be chosen. To support this position, they cited the immensely successful canal networks in Britain and Holland, which had demonstrated so clearly that riverbeds were treacherous and even dangerous for internal navigation. They did not have to look to Europe to prove their point. Right in their own backyard, the Mohawk looked like an ideal east-west water route, and yet it had stubbornly fought off the costly efforts of the Western Company to tame it into full navigability.

* * *

When the commission's report finally appeared in March 1811, public response was divided. There were those who were proud that a young nation could undertake a project of such imposing size. Doubters continued to balk at the sheer size of the proposal, despite Jonas Platt's effort to discourage the skeptics by selecting some of the most famous and respected men in New York State for the commission. The most serious opposition loomed up among the settlers in the west, jealous of their thriving trade with Montreal. Joseph Ellicott, the Holland Land Company's chief representative in the west, now wrote to his superior in Philadelphia, "I have therefore much less opinion of this great object than I formerly had. . . . I am pursuaded [*sic*] that it will be more advantage at once to make Montreal our market; and it makes no difference to us what Market we go to; the great object is to go to such a Place where we can make the most profits."[29]

These were all words. The action was still in the legislature, and here the momentum was positive. A month after the publication of the report, the legislature passed the first of a long series of canal laws it would enact over the years ahead. This bill added the distinguished names of Robert Fulton and Robert Livingston, his business and engineering associate, to the board of commissioners. It appropriated $15,000 to finance the board's further activities. It authorized the commissioners to take all necessary steps to arrange for financing the canal, to purchase the interests of the Western Inland Lock Navigation Company, to seek grants of land from property owners along the proposed route, and, most important, to approach Congress for fulfillment of the splendid promises of Thomas Jefferson and Albert Gallatin.

A few weeks later, the commissioners met to allocate their assignments. The purchase of the Western Company's properties had to be carried out and land grants had to be negotiated. Morris was given the responsibility of exploring the probable cost of loans from private investors abroad as well as in the United States. Eddy and Fulton were to seek out the engineers who would design and supervise the construction of the canal. And the two leading commissioners, Morris and Clinton, faced the prospect of going to Washington to enlist aid from both the federal government and from the neighboring states expected to benefit from the prosperity the canal was certain to generate.

At long last, the gears seemed to be meshing. That was an illusion. Many obstacles lay just ahead.

CHAPTER 9

WAR AND PEACE

The next three years were like a roller coaster for the Erie Canal. Although De Witt Clinton and the canal's other supporters kept up the pressure in the face of stubborn political wrangling, the grim distractions of the War of 1812 nearly derailed the whole project. Like all wars, this one had its fair share of unintended consequences.

Clinton and Morris had a disastrous trip to Washington late in 1811. Even before they left New York, Clinton's local political enemies, completely ignoring Morris's strong Federalist sympathies, claimed the whole enterprise was just an electioneering jaunt for him. As one of them put it, "to construct a rail road from the earth to the moon could not be treated with more derision."[1]

When they reached Washington, President Madison appeared to be cooperative at first—perhaps because he had speculated in land in the Mohawk valley some years earlier. Then he started quibbling about constitutional scruples: without explicit mention of canals in the Constitution, did the federal government have the right to appropriate money for that purpose? The president finally did send a message to Congress, reminding them of the strategic importance of improvements to the national transportation network, but the legislators were totally unresponsive.

Contrary to Clinton and Morris's expectations, the states adjacent to New York also sat on their hands instead of providing support. The canal commissioners would later describe this behavior as "state jealousy, [operating] with baleful effect, though seldom and cautiously expressed."[2]

Finally, Secretary of the Treasury Gallatin turned out to be as unyielding as Congress. Gallatin claimed he was enthusiastic about the Erie project, but the prospect of imminent hostilities with Britain was driving the Treasury rapidly into the red. He could no longer boast, as he had in his report three years earlier, that the government's resources were "amply sufficient for the completion of every practical improvement." In fact, the surplus of $6.3 million in 1811—the third largest in the twenty-two year history of the United States—would swing to a deficit of $10.5 million in 1812 and peak at a deficit of $23.5 million in 1814 as the war forced government spending to more than quadruple.[3]

Morris and Clinton had not asked Gallatin for money! Fully aware of the sad state of public finances, they had suggested instead grants of federal land to all the states involved, especially in Indian territories to the west, which New York and the other states could then sell to raise money. The answer was still no.

* * *

The New Yorkers had had a surfeit of hopes betrayed in the nation's capital. The canal commissioners made no effort to disguise their wrath at the duplicity they had encountered in Washington. "It remains to be proved," they observed in a report dated March 1812, written in large part by Gouverneur Morris, "whether they judge justly who judge so meanly of our councils."[4] Now they took the cooperative case Morris and Clinton had offered in Washington and turned it completely on its head. As the commissioners described their revised objective, "*now* sound policy demanded that the canal should be made by the State of

New York and for her own account." Any accrual to others of the manifold benefits to be delivered by the Erie Canal would be incidental.*

Despite the truculent tone of this statement, the State of New York had long experience financing projects "for her own account" as well as financing and directly encouraging enterprises in the private sector. The citizens of the state welcomed state intervention into these activities, without any misgivings about laissez-faire or government interference in private business affairs.

Aid to farmers in improving their productivity had been an interest of the state from the very beginning. In addition to authorizing start-up loans to entrepreneurs in manufacturing, the legislature had passed a general incorporation law in 1811. Seven years later, 129 charters had been granted to manufacturing firms, including in the charters such delightful perks as exemption from duty on juries or in the militia for owners of companies producing goods of cotton, linen, and wool. Manufacturers of machinery had it even better: they also were free from seizure for payment of debts.[5]

In addition to declaring the economic independence of New York State, the March 1812 report reviewed the other major matters the commissioners had considered in the first report. They declared—with no recorded dissent by Gouverneur Morris—that the inclined plane was to be just one of several methods under consideration for the design of the artificial waterway across the state. On the basis of further investigations and additional surveys, primarily by James Geddes, the commission raised its estimate of the total cost of the canal from $5 million to $6 million. The increase, they contended, was merely "a trifling weight" to the million people of New York State.[6] Now backed by an expert opinion from the English engineer William Weston, the commissioners reaf-

*As the years went by, and American industry and commercial trade developed, these views would change. By the 1830s, the voices for operation of public improvements by private industry became more strident, based on concerns about public corruption and expansion in public debts. See Nathan Miller, *The Enterprise of a Free People*, chapter 1, and Ronald Shaw, *Erie Water West*, pp. 308 and 398.

firmed the superiority of the interior route over the use of Lake Ontario. Weston had no doubt about the practicability of the project and was rhapsodic about the possibilities of the "noble and stupendous plan" laid before him by the commissioners. Its success, he predicted, "would baffle all conjecture to conceive"[7]

The arithmetic employed to justify an expenditure of $6 million was interesting. Based on an estimate of the increase in agricultural and other business activity that would develop along the route of the canal, the commissioners predicted westward traffic amounting to 250,000 tons within twenty years and a similar amount moving east, or 500,000 tons in all. At a tariff of $2.50 a ton, the canal could service its debt without any difficulty. Even if the traffic were half this estimate, the report continued, the total revenue of $600,000 would still take care of all the interest due on a 6 percent loan of as much as $10 million dollars.

These were solid projections. In his contribution to the Gallatin report, Robert Fulton had estimated that in 1807, based on data in customhouse books, 400,000 tons of freight were moving annually up and down the Hudson River. By 1818, when the canal had been under construction for a year, Charles Glidden Haines, Clinton's secretary at one point and an official of the New-York Corresponding Association for the Promotion of Internal Improvements (De Witt Clinton, president), reported that the cost of moving freight by land from Albany on the Hudson to Lake Erie was on the order of $100 a ton.[8] On the fair assumption that freight traffic would grow rapidly once the canal was in place—Fulton had projected 1 million tons a year—and considering Fulton's additional calculation that one horse pulling a canal boat could perform the work of forty horses pulling wagons on land, the commission's projections appear to be eminently reasonable. In fact, they would turn out to be too conservative by a wide margin.

The commissioners were, however, fully aware of the skepticism and local jealousies still clouding the future of the canal. Morris went to the limit in piling on the rhetoric:

> There can be no doubt that those microcosmic minds which, habitually occupied in the consideration of what is little, are incapable of discerning what is great . . . [and] will, not unsparingly, distribute the epithets absurd, ridiculous, chimerical, on the estimate of what [the canal] may produce. The commissioners must, nevertheless, have the hardihood to brave the sneers and sarcasms of men, who, with too much pride to study, and too much wit to think, undervalue what they do not understand, and condemn what they cannot comprehend. . . . And even when . . . our constitution shall be dissolved and our laws be lost . . . after a lapse of two thousand years, and the ravage of repeated revolutions, when the records of history shall have been obliterated . . . this national work shall remain. It will bear testimony to the genius, learning, the industry and intelligence of the present age.[9]

Three months later a special session of the legislature gave the recommendations of the commissioners the force of law. The commissioners were now authorized to purchase the rights of the Western Company, to accept land along the proposed route that might be contributed by individuals or business firms, to negotiate a loan of the necessary funds in the name of New York State, and to pursue surveys and other necessary work for the final design of the canal.

Good as all that sounded, the law was enacted within days of the declaration of war against Great Britain. The delay would be costly.

* * *

De Witt Clinton had many matters on his mind in addition to his intense concentration on transforming the canal from visions and meetings and expeditions into a functioning reality. Although Clinton's sixth term as mayor ran out in 1810, he was reappointed the next year and, at the same time, elected lieutenant governor of New York State—a position he took in preference to returning to the national Senate in order to sustain his influence in his main power base in Albany.[10]

The governor at the time was the curly-headed Daniel Tompkins. Tompkins cultivated the image of the humble farmer's boy from the start of his political career. He was fond of claiming, "There's not a drop of aristocratical or oligarchial blood in my veins." This was not an exaggeration. He had started out in life as a penniless student and was recruited early on by Tammany Hall. After stints in Congress and the New York Supreme Court he had been elected governor in 1807, with the backing of De Witt Clinton, and would hold the office for ten years.* But the friendship with Clinton would fall apart when Tompkins backed Madison against Clinton for the presidency in 1812.

Tompkins was also a strong patriot who contributed to the success of the wartime operations on the border between New York State and Canada, in part by extracting $30,000 from the federal government to help finance the defense of his state from incursions from Canada. He also pledged his personal credit when the New York banks refused to lend money on the security of the United States Treasury notes without his endorsement, and then he bought the weapons of private citizens to equip the New York State militia for the defense of the state (and had himself named commanding officer of the troops).[11]

Despite his attachment to New York, Tompkins yielded to the temptation to run on the Republican ticket as vice president of the United States under James Monroe. The first of his two terms, beginning in 1817, was much happier than the second, when he was caught up in a messy altercation over whether he had improperly handled the state's money during the War of 1812 or whether the state was indebted to him. The Federalists accused him of spending $120,000 of the government's money for personal use, but he could not rebut the accusation because he had kept such poor records. He then proceeded to produce a personal audit indicating that the government actually owed him $130,000. The controversy raged on, and the accusations broke his heart. His mind

*The states were where the action was: beside the examples of Clinton and Tompkins, John Jay had resigned from the U.S. Supreme Court in 1795 to run for governor of New York. He won.

began to fail him, and he died in June 1825, drunk and alone in his home on Staten Island.

* * *

At about the time he was elected lieutenant governor in 1811, Clinton and his wife moved from Cherry Street in New York City to Richmond Hill, a large estate just south of Greenwich Village. Richmond Hill had been occupied by John and Abigail Adams when John was vice president of the United States and subsequently by Aaron Burr in the same role. The Clintons now leased the property from John Jacob Astor. The property sloped down to the shores of the Hudson and was graced by a wide lawn with statues and a fine view of the New Jersey wheat fields. The spacious mansion was fronted by a two-story portico in the Greek style. Clinton also moved his working headquarters in 1811: by November, he was receiving visitors in the elegant new City Hall, which still stands at its original location.

On March 11, 1812, the canal commissioners had submitted their report, much of it in Clinton's hand, to the legislature. His beloved uncle, George, the vice president of the United States, died on April 20—and was eulogized at the funeral by Gouverneur Morris. His father, James Clinton, the Revolutionary War general, would pass away in December. On May 28, although both Clinton and Madison were officially members of the Jeffersonian Republican Party, Clinton was nominated for president by the Federalist convention in New York, which then recommended him to the rest of the party.

Gouverneur Morris, deeply concerned over the war spirit in Washington, had invited Clinton early in May 1812 to meet with him at the Morris ancestral home of Morrisania in the Bronx, ostensibly to talk about George Clinton but in fact to thrash out together what might be done about the current national political scene. Morris was for once a pessimist. "In the degenerate state to which democracy never fails to reduce a nation," he observed, "it is almost impossible for a good man to govern, even could he get into power, or for a bad man to govern well."[12] Morris wanted to call for a convention of northern states, where the opposition to war was strongest, largely because a war would interrupt the profitable trade those states carried on with Britain. Furthermore, like Clinton, Morris was concerned that his nation of fewer than eight million people was poorly prepared for hostilities with a great power.

Morris was a Federalist and a long way from favoring "the common man" in the Jeffersonian fashion of the Republicans. Despite his strong feelings on these matters, Morris would throw his full support to the Republican Clinton in the presidential campaign. It was a sign of the times. After three Republican terms in the presidency—Jefferson's two terms and one term of Madison's—the Federalist Party was no longer the force it had been when Washington and then John Adams had led the nation. If the Federalists had any hope of defeating Madison, their best chance was to be led by a renegade Republican on a splinter ticket. What remained of the Federalists would be likely to provide whatever support they could to an opponent to Madison. This was by no means the only time that Clinton would be forced to rely on Federalist support because of defections from his side by fellow Republicans.

Less than a month after Morris and Clinton conferred in Morrisania, Madison asked Congress to declare war on Great Britain. By 1812, the continuing British harassment of U.S. ships had taken a terrible toll on American commerce. Total exports in 1810 were $67 million, with more than two-thirds going to Europe. By the time Congress declared war in June 1812, exports had fallen by 40 percent and shipments to Europe had declined by a half. Matters would only grow worse. In 1814, total exports were a mere $7 million, of which only $1 million went to Europe.[13] The economic impact of the war was not limited just to trade in goods and services. Americans' prime source of capital was cut off. As the French economist Michael Chevalier described the situation twenty years later, with some degree of hyperbole, "Bankruptcy smote them like a destroying angel, sparing not a family."[14]

Clinton had been outspoken in his opposition to war because he was convinced the country was not yet prepared for combat with a great power like Britain. But he was sufficiently infected by the political bug in his run for the presidency to talk out of both sides of his mouth, depending upon whether the crowds he addressed were largely hawks or doves. The Federalists' valid claim that Clinton's "herculean" mind was "enlightened by extensive erudition" counted for little when he was running against Madison, a man whose intellectual brilliance was widely recognized and beyond dispute.

Gouverneur Morris, in contrast, was unwavering in his opposition to the war and in his disdain for the president. He claimed to have been informed that Madison never went to bed sober, although "whether intoxicated by opium or wine was not said."[15] Morris saw the war as an excuse to enrich the agricultural base of the slave states and inhibit the increasing industrialization in the north. In a prescient and economically sophisticated observation, in 1812, Morris sensed the growing split between north and south: "Time . . . seems about to disclose the awful secret that commerce and domestic slavery are mortal foes; and, bound together, one must destroy the other."[16]

Clinton lost the election to Madison by 39 electoral votes, with not a single vote from south of the Potomac River. Pennsylvania's 25 electoral votes were in doubt for a whole month as the recounts went on and on and finally ended up in Madison's favor, even though Clinton's running mate, Jared Ingersoll, was attorney general of that state. The Federalist Party never regained its national strength after this defeat. Morris was especially bitter about Pennsylvania, which "may be led to cover with her broad shield the slave-holding states: which, so protected, may for a dozen or fifteen years exercise the privilege of strangling commerce, whipping Negroes, and bawling about the inborn inalienable rights of man."[17]

* * *

Once war breaks out, the military spirit is hard to resist.[18] Clinton now managed to have himself appointed a major general, but political enemies kept him away from any active involvement. He had plenty to do to support what came to be known as "Mr. Madison's war" in his regular job as mayor of New York City, supervising the construction of armed camps in Brooklyn and Harlem Heights as well as a fort at Hell's Gate, in addition to a search for financing to support the soldiers at those camps. All those sites had been battlegrounds leading to American defeats in the earliest stages of the Revolution.

His fellow canal commissioner Peter Porter, who had been craving an excuse to invade Canada, got himself appointed major general of the New York Volunteers even though, like Clinton, he had no military experience. He proceeded to raise his own untrained troops to join in the fighting. Unlike Clinton, he and his troops saw action in several battles, and a joint resolution of Congress dated November 3, 1814, awarded Porter a gold medal with a citation for "gallantry and good conduct in the several conflicts of Chippewa, Niagara, and Erie."[19]

* * *

Madison's war message to Congress on June 1 followed convention in portraying his own country as the injured party. He focused on Britain's "lawless violence" on the seas over so long a period of time. Then he drew Congress's attention to "the warfare just renewed by the savages on one of our extensive frontiers, a warfare which . . . spare[s] neither age nor sex and to be distinguished by features particularly shocking to humanity." As Madison summed up his view of the struggle, "We behold, in fine, on the side of Great Britain a state of war against the United States, and on the side of the United States a state of peace toward Great Britain."

Well aware of the constitutional constraints on the president in declaring war, Madison reminded Congress that the choice between accepting "these progressive usurpations" or fighting to defend the nation's rights was a matter for the legislative branch to decide. He nevertheless pronounced himself assured that the decision would be "worthy of the enlightened and patriotic councils of a virtuous, a free, and a powerful nation."

On June 4, the House of Representatives obliged by declaring war on Great Britain by a vote of 79–49, with almost all the New York State representatives voting against. The debate in the Senate was both more lengthy and more intense. The vote there did not come until June 18, and the margin for war was slim. In one of the great ironies of history, the British foreign minister announced just two days earlier that Britain would relax its blockade on American shipping—but that news would not reach the shores of the United States for another five weeks, by which time the opposing armies were fully engaged.

* * *

The early stages of the war were a demoralizing series of defeats on home territory, including the surrender of Detroit soon after the outbreak of hostilities. Too much time had passed since the Revolution, when many officers in the Continental Army already had combat experience fighting on the British side in the French and Indian Wars. Now the American land forces had no battle-hardened troops or commanders—Jefferson remarked he had never seen "so wretched a succession of generals."[20] Henry Clay observed that Madison "is not fit for the rough and rude blasts which the conflicts of nations generate."[21] Meanwhile, the British had been at war for half a century, with few interludes of peace.

Popular support for this war was far weaker than in the Revolution. The Senate had dickered through two weeks of secret sessions before voting in favor of the war resolution, and even then the vote was close, 18–13. Opposition was intense in New England, where the business community was deeply—and justifiably—concerned about total loss of trade with Europe. Indeed, only 108 vessels entered the United States during the entire year of 1812.[22] The governors of Massachusetts, Connecticut, and Rhode Island at first refused to call up the state militias to go into combat. The New England banks, the most prosperous in the nation, even refused to lend the government money for the war effort, forcing Secretary of the Treasury Gallatin to go hat in hand to smaller banks, which then took advantage of the situation by charging usurious rates of interest.[23]*

But on the west side of the country, sentiment was strongly in favor of the war, especially among New Yorkers. Many westerners saw the war as a perfect excuse for invading Canada, with its rich forests, fur sources, and agricultural lands. Although Jefferson, for one, expected a Canadian invasion to be little more than a matter of marching,[24] others felt differently. When war was declared, fear of invasion *from* Canada prompted the settlers out in Holland Land Company country to move back eastward "in droves," ignoring Resident Agent Joseph Ellicott's soothing assurances and entreaties to remain.[25]

They were right to run. The fighting in the west turned out to be much bloodier than a matter of marching. The first effort by the Americans to invade Canada was a disaster. After Detroit fell so early in the conflict, repeated invasion efforts by both sides led to casualties with no permanent victories for either contender and shockingly stupid military leadership on the American side.[26]

Throughout the whole struggle, the Americans had to fight simultaneously on both the western and the eastern sides of Lake Erie. Gory battles around Niagara Falls, on the Canadian and on the American sides, would resolve nothing. Many of the western New Yorkers returned to their homes in the course of 1813. The Americans themselves burned Buffalo in a December snowstorm to deny the British any hope of wintering in the vicinity—while inadvertently leaving barracks and tents for

*New England would even threaten to break up the union. In October 1814, delegates from the New England states gathered in Hartford, where they seriously considered secession and proposed constitutional amendments to redress what they considered unfair advantages held by the South under the Constitution.

1500 men in perfect condition, along with supplies of ammunition. Even Ellicott at that moment fled from his headquarters in Batavia, citing the atrocities of the British and their Indian allies in killing women and children as well as the men and the resulting "terror, consternation, and dismay" among local citizens.[27]

Most of the American victories were scored at sea against the vaunted British navy, even though the American oceangoing navy was vastly outnumbered by Britain's. The British had 245 frigates, each carrying thirty to fifty guns, and 191 ships of the line with sixty to eighty guns.* The Americans had no ships of the line and only 7 frigates, one of which would see no action. The ingenious design of the American frigates made all the difference. These marvels of naval technology, which dated back to John Adams's time as president, could outmaneuver Britain's massive ships of the line and also carried more guns than British frigates.[28] The combination of greater firepower and greater maneuverability kept the British on the run. The most famous of these frigates was the *Constitution*, or "Old Ironsides," which won two skirmishes in the first months of the war against British frigates.

The American navy also achieved crucial victories in battles on the waters of Lake Erie, coveted by both parties because of its access to Canada, to the other Great Lakes, and to the Mississippi beyond. The New Yorkers were especially concerned because the lake was also the planned western terminal of their proposed canal. As neither side had enough warships to mount a proper battle on the lakes at the outset, the real struggle took the form of a shipbuilding race along the shores of the lake. This was no easy task for the Americans, with limited labor supply in the area, no sawmills and no factories in New York, and no good transportation system for bringing in the necessary supplies. The facilities at Erie, Pennsylvania, on the shore of Lake Erie about a hundred miles west of Buffalo, did manage to supply the fighting forces in the

*The expression "ship of the line" refers to a warship with at least two gundecks and designed to have a place in the line of battle.

lake with shipwrights, blacksmiths, caulkers, and common labor, as well as canvas, rigging, cannon shot, and anchors—but no cannon. The cannon had to come all the way by road from the east coast.

Commodore Oliver Hazard Perry, then only twenty-eight years old, took command at Erie of American forces on the lake in March 1813. A veteran of fighting pirates in the Mediterranean and the Atlantic, Perry managed to build a fleet of six ships by July. He had also recruited volunteers to man them—including freed blacks, farmers, and soldiers from the army. Perry named his flagship the *Lawrence* after his best friend, James Lawrence, who had recently been killed in naval combat, and then he hoisted a flag painted with Lawrence's enduring words "Don't give up the ship!" Now he was eager for battle.

On September 10, joined by a few additional small craft from Buffalo, Perry engaged a British squadron on the lake. The battle began grimly when Perry did what his friend had not: he gave up his ship as concentrated British fire killed or wounded 80 percent of his crew. He was not to be deterred, however. Carrying the precious flag, he transferred on a small boat to another ship, sailed right into the British line, and compelled surrender within fifteen minutes.

The lake was now secure for the Americans. Perry then doubled his claim to immortality. His message of victory to General William Henry Harrison proclaimed, "We have met the enemy and they are ours: two brigs, one schooner, and one sloop."

* * *

The prospects for the Erie Canal seemed to grow dimmer as the war progressed. The canal commissioners could only bide their time as hostilities proceeded. Perhaps motivated by Weston's reassurance that "from your perspicuous topographical description, and neat plan and profile of the route of the contemplated canal, I entertain little doubt of the practicability of the measure," the commissioners sent the legislature a new progress report in March 1814. They confirmed the findings of their earlier messages, added an elaborate set of estimates of transport costs, recommended appropriate tolls to be charged, and provided detailed comparisons of shipping costs in other parts of the country, especially along the Mississippi.

This reminder that the canal was still a matter of high importance had an unexpected result. The report ignited the opposition, who had gathered strength from the dissipation of energy among the canal's supporters under the clouds of war. In particular, resistance was building up among farmers living far away from the route of the canal, who were concerned their taxes would only finance competition in their markets. A month after the publication of the 1814 report, the opposition forces in the legislature inserted a clause in a supply bill that annulled the authority to create a fund for financing the construction work on the canal. As De Witt Clinton later described this underhanded maneuver, "The commissioners were thus frittered down into a board of consideration . . . without power and without money."[29]

"The Canal bubble it appears, has at length exploded," reported Joseph Ellicott to his superior at the Holland Land Company in Philadelphia.[30] In 1815, the commissioners did not even bother to make a report to the legislature. That year, Clinton's opponents in the legislature removed him as mayor of New York City; he had already lost the lieutenant governorship in 1813. Thomas Eddy described the friends of the project as "entirely discouraged, and [having] given up all hopes of the legislature being induced again to take up the subject, or to adopt any measure to prosecute the scheme."[31]

* * *

But the most important friend of the project was far from discouraged. In fact, Clinton's exclusion from political office was a blessing in disguise for the canal. Free of routine responsibilities, he was able to focus the full force of his abilities on promoting the canal, and he took up the challenge with gusto. As far as he was concerned, the case for the canal was stronger than ever and would prove impossible to suppress once peace was declared.

The widespread fighting in the west had confirmed the area's military and economic importance. As suspicion of Canadian intentions lingered on, there was a new urgency to avoid Lake Ontario for east-west traffic. Most important, the war revealed the tragic lack of adequate transportation facilities to link the lands around Lake Erie with the east. Supply shortages constantly hampered effective military operations. Wagons repeatedly broke down on the rutty roads, and men as well as horses were exhausted by the hardships of travel over the long distances. A report issued in 1816 by General James Tallmadge, a congressman from New York, claimed that east-west transportation costs for armaments during the war were so high that a cannon manufactured at an eastern factory for $400 cost from $1500 to $2000 to ship to Lake Erie, while a barrel of pork for the troops on the western frontier ended up costing the government $126—and all that without counting in the plodding pace of horses and wagons "over roads so abominable as to make cannon balls cost a dollar a pound."[32]

Another and more sophisticated theme came to the fore in the years after the war: the importance of developing a home market in the face of uncertain economic conditions among America's customers in Europe. The economic case for concentrating on American customers was clear enough. Postwar deflation abroad influenced postwar deflation at home, as prices collapsed even before the official end of hostilities. By 1821, prices were down 40 percent from the 1814 peak. American exports, after a brief upswing, fell off sharply as the slump in European economic activity gathered momentum. Meanwhile British manufacturers, deprived of the insatiable wartime market of their government, now turned aggressively to exports, especially to their former enemies across the seas.

Although the development of the home market appeared as a natural priority under these conditions, the appeal ran deeper than the economics involved. Americans were proclaiming to the world that their United States was destined to become a first-class power—a motif they would repeat on many occasions over the next hundred years. Adam Smith himself had declared in 1766 that "good roads and canals and navigable rivers, by diminishing the expense of carriage, put the remote parts of the country more nearly upon a level with these in the neighborhood of large towns; and that they account for the greatest of all improvements."[33]

After emphasizing how waterways create networks between provinces and districts of the same country, Clinton's protégé Charles Glidden Haines declared in an 1818 pamphlet that "those nations who have been destitute of means of inland navigation, either by rivers or canals, have remained from one age to another in the same barbarous and uncivilized state. . . . Such intercourse is vitally essential to the welfare of nations."[34] Haines becomes even more rhapsodic as he expands his case: "It is evident that no country in the world, ever presented natural advantages for internal trade and canal navigation, so bold, so noble, so striking as our own."[35]

* * *

Almost a hundred years had passed since Cadwallader Colden had asked, "How is it possible that the traders of New-York should neglect so considerable and beneficial trade for so long a time?" Little did Colden know just how possible it was. In his wildest imagination, he could not have imagined that, quite aside from two wars with the British, it would take nine decades of successive explorations, passionate speeches, eloquent pamphlets, authoritative surveys, distinguished commissions, and elaborate reports to overcome the potent combination of vested interests and timidity in the face of bold vision.

But now the barricades to progress were ready to crumble. This particular attack on the opposition began in familiar fashion: one day in the autumn of 1815, Thomas Eddy "could not . . . resign a favorite project" and invited his old friend Jonas Platt to breakfast. Eddy was convinced they must make one more effort to make the canal a reality. He proposed to Platt that they should reverse the strategy they worked out together back in 1810. This time around, the undertaking should begin by arousing the public instead of going directly to the lawmakers. To that end, he proposed organizing a mass meeting at the City Hotel in New York, "to urge the propriety and policy of offering a memorial to the legislature, pressing them to prosecute the canal from Erie to the Hudson."[36] Platt was enthusiastic and urged Eddy to get going. Eddy once again began with De Witt Clinton, who went to work on the project immediately, attracting additional speakers and generating the necessary publicity to attract a big crowd.

The meeting, held on December 3, 1815, was a huge success, with an overflow audience. A committee was appointed to draft a "memorial" addressed to the legislature, almost all of which was the handiwork of De Witt Clinton. Armed with Clinton's voluminous and persuasive words, Eddy and his colleagues then proceeded to organize additional mass meetings throughout the city and in twenty-five other cities throughout the state, gathering thousands of signatures to be delivered to the legislature in support of Clinton's Memorial.*

The Memorial runs to thirteen tightly packed printed pages. Although Clinton includes important passages of his powerful rhetoric, he is equally unremitting in the provision of detail and data. He evokes the success of canals in contributing to the greatness of ancient Egypt and China as well as to modern Holland and England, but he does not linger over the glories to come from uniting "our Mediterranean with the ocean," nor does he stop to elaborate on how "the wilderness and the solitary place will become glad, and the desert will rejoice and blossom as the rose." All that is taken for granted. Rather, Clinton explains at

*In his *History of the Canal System of the State of New York*, Noble E. Whitford remarks, "This agitation brought before the Legislature an appeal from more than one hundred thousand petitioners to proceed at once with the work of making a canal." The population of New York State at that time was about one million.

length the superiority of New York over its competitors as an artery to the west, the most desirable route for the canal, plans for its design, the remarkable variety of crops and merchandise that the canal will transport, itemized estimates of costs and tolls based on the Gallatin report—as well as the construction costs of other canals, ranging from the Canal du Midi in southern France to the Middlesex Canal in Massachusetts—and the ease of finance and the large acreage of lands that will be contributed by the Holland Land Company and other landholders seeking public favors.

The war, he contends, had proven beyond question the need for the canal and emphasized the "importance of this communication [via New York State] . . . through the most fertile country in the universe." Time is of the essence. Lands and men left idle in the wake of the war are now readily available at low cost. Delay would excite "injurious speculation" on the one hand and, on the other, give the opposition opportunity to regroup.

The home market is an urgent matter: "Our merchants should not be robbed of their legitimate profits . . . public revenues should not be seriously impaired by dishonest smuggling, and . . . the commerce of our cities should not be supplanted by the mercantile establishments of foreign countries." Then, on a strikingly modern supply-side note, Clinton goes on to argue that the canal will raise the value of "national domains," thereby facilitating the repayment of the national debt and releasing resources "to be expended in great public improvements; in encouraging the arts and sciences; in patronizing the operations of industry; in fostering the inventions of genius, and in diffusing the blessing of knowledge. . . . [The canal will] convey more riches on its waters than any other canal in the world." With his usual remarkable foresight, Clinton asserts that, in addition to the great good the canal would do for the state as a whole, it would transform New York City into "the great depot and warehouse of the western world."

Clinton does allow himself to introduce his soaring rhetoric in the final paragraphs. He bases his final arguments on the same theme Washington had struck over and over: the critical importance to the nation—not just New York State—of trade and commerce in binding the west to the Atlantic states. "However serious the fears which have been entertained of a dismemberment of the Union by collisions between the north and the south," Clinton begins, displaying his weird sense of what the future might hold, "the most imminent danger lies in another direction. [A] line of separation may be eventually drawn between the Atlantic and the western states, unless they are cemented by a common, an ever acting and a powerful interest." Still echoing Washington's views, Clinton contends that the canal, by providing the channel between "the commerce of the ocean, and the trade of the Lakes, . . . will form an imperishable cement of connexion, and an indissoluble bond of union." New York State, "standing on this exalted eminence," is both Atlantic and western and therefore "the only state . . . in which this great centripetal power can be energetically applied."

The case is beyond dispute: "Delays are the refuge of weak minds, and to procrastinate on this occasion is to show a culpable inattention to the bounties of nature; a total insensibility to the blessings of Providence, and an inexcusable neglect of the interests of society. . . . The overflowing blessings from this great fountain of public good and national abundance, will be as extensive as our country and as durable as time. . . . It remains for a free state to create a new era in history, and to erect a work more stupendous, more magnificent, and more beneficial than has hitherto been achieved by the human race."[37]

CHAPTER 10

THE SHOWER OF GOLD

Construction on the Erie Canal began on the Fourth of July 1817, just three days after De Witt Clinton had been elected for his first term as governor of New York State. Suddenly, all the years of debate, doubt, and division seemed to melt away. It was such a great moment that even former enemies would reverse course and transform themselves into champions of the canal. Looking back, people could only wonder why the struggle had been so protracted.

There had been good reason.

At first glance, the resiliency of political resistance is more understandable than the persistence of disbelief and skepticism. Virginians-against-Yorkers reached all the way back to Washington's day and persisted right up to Madison's. The neighboring states were jealous of New York's strategic location and God-given landscapes. Even in New York State, there were many farmers concerned that the canal would open New York to the production of farms far more fertile and productive than their own—and therefore able to offer their output at lower prices.

The citizens of New York City in particular had many misgivings, despite the rosy forecasts the canal enthusiasts kept repeating about the city's future. Aside from the normally timid people scared off by the financial disasters of the Western Inland Lock Navigation Company, the city's opposition stemmed primarily from merchants and manufacturers concerned over new taxes for which they would receive an insufficient return from a greater volume of business. Assemblyman Peter Sharpe warned that taxes were already draining the city to the limit and that the "most respectable and opulent of [the city's] merchants are daily becoming bankrupts. . . . [The state] will sink under [the debt whose] magnitude is beyond what has ever been accomplished by any nation." In the state Senate Peter Livingston declared that only a madman or a fool would proceed with such a wildly unrealistic project: "From diggers on up," he predicted, there were "no worse managers of funds than the public."[1] Meanwhile, infighting between Clintonians and anti-Clintonians had developed into an indoor sport, and frequently an outdoor sport as well.

But indefatigable political forces like these do not exist in a vacuum. If the hostility to the canal had amounted to nothing more than jealousy or bald self-interest, the glittering future promised by its supporters would have quickly conquered the opposition. The opposition's case had more substance than that. Disbelief was real, not a blind for political discord: many people were simply unable to visualize how such a novel, gigantic, and hugely expensive project could ever fulfill those glowing promises. When Thomas Eddy and Jonas Platt launched their campaign for credibility in 1810, they were well aware of the educational challenge before them. They expected it to be an even tougher obstacle than the ongoing political struggles.

The few canals already completed in the United States were less than fifty miles in length—many much less than that—and traveled through populated countrysides without great waterfalls or deep valleys. There were only two precedents for a project of this magnitude: the intricate canal network the English had built over the past fifty years to link their burgeoning industrial centers throughout the Midlands, and the Canal du Midi connecting the Atlantic to the Mediterranean in southern France. Neither bore any resemblance to the projected waterway across New York State. Both were located in areas where economic activity was already well established, where population densities were many times larger than in the country west of the Hudson River, where distances between towns were short, and where the shape of the land to be crossed was obliging. The King of France had been a powerful sponsor for the 115-mile Canal du Midi, and emerging industrialists like Matthew Boulton and Josiah Wedgwood in Britain represented a wealthy and rapidly growing pool of financial capital. The model for the British canals, the Bridgewater Canal built in 1759, ran all of 7 miles toward Manchester, one of the most populous and dynamic cities in the whole country. None of the English canals, independent of branches, extended more than 100 miles.

The Erie Canal had none of these initial advantages. As Noble Whitford, the most meticulous of the many biographers of the Erie Canal, put it in 1905, "When we recognize the primitive conditions and review the difficulties, we do not wonder that the people of a struggling republic stood aghast at the vast enterprise." And he goes on to cite those who predicted that in Clinton's "big ditch would be buried the treasure of the State, to be watered by the tears of posterity."[2] One of Whitford's contemporaries, the writer Samuel Hopkins Adams, would echo these words in 1944 in a riotously funny novel about the canal, in which at one point a character jeers, "The canawl! The canawl! I'll spit you all the canawl you'll get." He spits copiously in front of his companions and then sings:

> "Clinton, the federal son-of-a-bitch,
> Taxes our dollars to build him a ditch."[3]

The proposed canal would extend 363 miles through what was then almost entirely wilderness. The population west of the Hudson River and beyond the Albany area had been less than 25,000 when Jesse Hawley published his essays in 1807.* Although almost 100,000 people were living there by 1817, that was still only 7 percent of the entire population of New York State.† Birmingham, England, just one major city in the center of the Midlands, had a population of about 70,000 people at that time.[4] Civil engineering as a profession did not then exist in the United States, but the engineers would have to confront abrupt and dramatic changes in elevation at many points along the way, such as Cohoes Falls at the east end and an even more difficult situation at the Niagara Falls escarpment at the west. Finally, despite New York State's cavalier attitude about it, the money involved was as intimidating as the cataract at Cohoes Falls. According to one authority, the proposed $6 million was equal to nearly a third of all the banking and insurance capital in the state.[5] Six million dollars would be the equivalent of about $90 million in today's purchasing power, but in an economy that was only a tiny fraction of the gigantic U.S. economy of the twenty-first century.‡

The word "visionary," either stand-alone or as an adjective, came to have a pejorative meaning when applied to canal enthusiasts. According to Jonas Platt, "hallucination" was the "mildest epithet" the opposition liked to employ.[6] The whole thing just seemed too good to be true. Without constant repetition of the accumulated results of successive explo-

*Similar hurdles faced the development of the American railroads in their early days. As late as 1874, the *Times* of London would describe the American railroads as running from "Nowhere-in-Particular to Nowhere-at-All." The financial historian Carter Goodrich has pointed out that this accusation was only half true. In the typical case—and the Erie Canal was typical in this instance—the project started somewhere and "ended at a point which was perhaps nowhere when the project started, but which was to become an important as a result of the improvement itself." See Goodrich, *Government Promotion of American Canals and Railroads*, p. 10 and fn. 18.

†The census of 1810 showed 23,416 people in western New York. The census of 1820 showed 108,981. As there were no interim counts, we do not know the exact number for 1817.

‡Rough estimates suggest that nominal gross domestic product in 1815 was in the area of $700,000,000, or about $8.5 billion in today's money; today's GDP is in the area of $10 trillion. I have derived these estimates based upon data on the War of 1812 in William D. Nordhaus, *The Economic Consequences of a War in Iraq*. Nordhaus reports that the War of 1812, stretching over three years, cost $90 million, so it was much more costly than the Erie Canal.

rations and analyses over the years, and without the many eloquent recitals of the rewards that would justify the effort and the risks, the canal would never have come into being. De Witt Clinton's powerful Memorial of December 1815 was the climax of this educational process, but his meticulously itemized citation of numbers and his elegant reasoning would have been just one more document if it had not been preceded by the glut of dreamers and madmen over the preceding twenty years.

* * *

There were still a few jolts remaining before the vision would finally meld into reality. After the electrifying response to the Memorial, the opposition could put up only a delaying action. They managed to block the canal for another eighteen months.

Governor Daniel Tompkins, the "humble farmer's boy," posed one of the more difficult obstacles. Following the publication of Clinton's Memorial, Governor Tompkins was eager to gain support from the westerners in Holland Land Company country, which would have made him pro-canal. On the other hand, Peter Porter was one of Tompkins's most powerful supporters in the west, and Porter remained adamantly in favor of the Ontario route. Tompkins chose to resolve this dilemma by equivocating. In a speech on February 2, 1816, for example, he began with enthusiastic support for road building; then he turned to the canal and proceeded to talk out of both sides of his mouth.

Clinton could not contain his wrath. He had already characterized Porter as "a tool and a dupe" and, in super-Clintonian language, he depicted Tompkins as "destitute of literature, science, and magnanimity—a mere creature of accident and chance, without an iota of real greatness."[7] When Tompkins subsequently recommended to the legislature that they employ the inmates of state prisons in erecting fortification, repairing the roads, or in constructing canals, Clinton lost all patience. "If this is not a full exposure of the cloven foot of hostility, I know not what is," he declaims, accusing Tompkins of "a sneer of contempt."[8]

Tompkins did not back down easily, even in the face of a lopsided vote of 91–18 in the Assembly giving the canal an official go-ahead. The bill stipulated that construction should begin along the relatively flat country between Rome on the Mohawk River and the Seneca River, a stretch of almost ninety miles centered on Syracuse. The bill also provided for financing this work by a tax on the lands bordering the canal by twenty-five miles on each side, plus the authority to borrow up to $2 million. But it was not to be.

When the bill came before the Senate, the leader of the opposition was Tompkins's ally, Martin Van Buren, who would in time be known as the Red Fox of Kinderhook (after his hometown, a village south of Albany) or the Little Fox. Whatever handicaps his five-foot-two size bestowed, Van Buren overcame them by always standing erect, impeccable, and fastidiously dressed. The son of a farmer who was also a tavern keeper, Van Buren disguised his determination and ambition with great charm and amiability. He was a hard man to put down.

On this occasion, he complained that the plans of the canal commissioners lacked sufficient detail about the final route of the canal, as well as its projected construction and financing, for the legislature to approve such a large project. Van Buren exhorted his fellow senators to strike out the authorization for construction in the Assembly's bill. They promptly obliged him, voting 19–6 to delay the work and calling for a brand-new set of surveys to support the detailed delineation of the proposed route. As a consolation prize, the Senate added a provision for $20,000 to pay for the additional research.*

The pro-canal forces in the Assembly made a final effort to rekindle their earlier victory but ultimately bowed to the wishes of the Senate on the last day of the session. The Tompkins forces had performed the neat trick of appearing to support the increasingly popular canal project while at the same time seeming more responsible than the impatient

*$20,000 in 1816 is the equivalent of about $300,000 in current purchasing power.

members of the Clintonian camp.[9] As James Geddes wrote to Clinton, "After all the expectations raised . . . it is still 'hope deferred.'"[10]

There was one more disappointment to come, and it was a familiar one—in Washington, D.C. The only surprise was that the New Yorkers actually expected a favorable outcome from Washington. Something did seem to be stirring early in 1816 when the U.S. Senate authorized the printing of 800 more copies of the Gallatin report of 1808, nearly as many as the 1200 printed originally. Then Representatives John Calhoun of South Carolina and Henry Clay of Kentucky sponsored a bill very much in the spirit of Gallatin's work, establishing a fund for internal improvements that specifically provided $90,000 a year for twenty years for a canal across New York State.

It is ironic that Calhoun, a leader in the fatal move to secession in the 1860s, could at this moment call for binding the Republic "with a perfect system of roads and canals. . . . Neglecting them, we permit a low, sordid, selfish and sectional spirit to take possession. . . . We will divide, and in its consequences, will follow misery and despotism."[11]

The Clay-Calhoun bill passed in the House by a razor-thin margin of two votes. Then victory eluded the New Yorkers once more, for Madison vetoed the bill. The president was still uttering concerns about the Constitution's silence on any kind of authorization for federal financing of roads and "watercourses." Was he so in Jefferson's shadow that he had forgotten the concluding phrases of the Gallatin report, "The National Legislature alone . . . superior to every local consideration, is competent to the selection of such national objects"? Madison's words also made a strange contrast to his deeds. On the same day he vetoed the canal legislation, he signed an appropriation bill of $100,000 for work on a road heading westward from Cumberland, Maryland (near the terminal of Washington's Patowmack Canal), to the Ohio River, a project authorized in 1805 during Jefferson's administration and now nearing completion.

Clinton's temper boiled over once again. He described Madison's veto as "a miserable sophism, contradicted by the whole course of his official conduct. . . . Lighthouses have been erected on the great lakes for the convenience and safety of inland trade. Are not canals equally conducive to the promotion of this important object? . . . Whatever gloss may be thrown over this reprehensible conduct of Mr. Madison, it requires no slight exercise of charity, not to connect it with jealousy of the growing prosperity of New-York."[12]

Clinton need not have been so exercised. Madison's veto was the last stand of the canal's opponents.

The New York legislature's canal bill of February 1816 contained positive features even as it postponed construction. In addition to the generous sum of $20,000 to finance the surveys, the bill appointed five new commissioners to replace the original group of seven. The legislature also directed the commissioners to come up with specific recommendations for raising the proposed $6 million to finance the canal. In order to link the Erie system to New England, the legislation instructed the commissioners to present plans for an additional canal running due north twenty-two miles from Albany to Lake Champlain, which forms the border between northern New York and Vermont. The Champlain Canal would be in effect a northern extension of the Hudson River.

The business at hand was no longer persuasion. Now the emphasis had shifted to the design, financing, and management of the actual construction of the canal. In selecting a new set of commissioners, therefore, the legislators focused on professional skills as well as political power—in contrast to the original group, where the balance of political alignments had dominated.

The five new commissioners included De Witt Clinton, Stephen Van Rensselaer, Joseph Ellicott, Myron Holley, and Samuel Young. Clinton was an obvious choice. Van Rensselaer offered huge wealth, first-class connections, and a fascination with science. He had spent long hours educating himself in the area, and was so devoted to it that in 1824 he founded an engineering school in his name, now known as Rensselaer Polytechnic Institute in Troy. Ellicott, the longtime resident agent of the

Holland Land Company and an able and experienced surveyor, was probably the best informed man in the state about the geographical features and available labor force in the west. To Clinton, Ellicott was also a welcome replacement for Peter Porter. Young was a scholar who would become an early exponent of laissez-faire economics; his vocal championship of free private enterprise over government in later years would lead him to consider "internal improvements . . . eternal taxation." At this stage, though, he was—like everyone else—still willing to accept public financing and management for a project as large as the Erie Canal.[13] Young's primary contribution was *A Treatise on Internal Navigation*, his authoritative survey of the canals in Britain and Holland. This lengthy work played a key role in educating the public and the politicians about the feasibility of the proposed canal. Holley, an assemblyman from the Finger Lakes region, was a staunch advocate of government-financed public improvements and an irrepressible promoter of the Erie Canal. Clinton was one of Holley's most devoted admirers because "he improved his mind by reading, reflection, and conversation [and] devoted his whole time and attention, mind and body, to the canal. . . . Some of the most luminous . . . communications have proceeded from his pen."[14]

But what about Gouverneur Morris, who had played such an important role as both inspiration and commissioner in years past? Morris's naïve insistence on the inclined plane was only the beginning of constant friction and conflict with his colleagues. When he was charged with responsibility for finding sources of finance for the canal, Morris succeeded in displaying his lack of practical good sense once again. After locating a Frenchman in Paris to solicit potential lenders, Morris proposed that this agent should receive the proceeds of the loans and then personally remit the funds to Albany. Morris's colleagues were incredulous that he would consider authorizing this individual to serve as principal rather than as an agent who would arrange for the lenders themselves to remit the money directly to New York. It might have been a different matter if Morris's choice had been a partner in a bank of some kind, but he was operating all by himself. There was a real risk the man could commit the state as borrower and then abscond, or might die, without having turned a penny over to the authorities in New York. Morris just did not get it. He was deeply offended at the abrupt rejection of the efforts he had put into this project.[15]

The final blow came when the other departing commissioners suggested to Morris that he prepare a draft of their valedictory report. Although his colleagues had gone along with Morris's first report in the interest of unity, this time around they could not accept his work. Morris would not accept their revisions. He refused to place his signature on the final version of the report and withdrew. He passed away later that year, at the age of sixty-four.

* * *

The five commissioners met in New York City on May 17, 1816, to respond to the legislature's demand for complete plans covering route, construction, and financing. They were well aware that Van Buren was waiting in the wings to dispute whatever their work would produce, which meant their report back to the legislature would have to be even more complete—and more compelling—than all of the previous studies together.

With De Witt Clinton presiding, their first decision was to divide the proposed Erie Canal—or the Great Western Canal, as they referred to it—into three great sections covering 363 miles in all: a downward-sloping western segment running 165 miles from Lake Erie to the Seneca River, a middle section with gentle rises and falls in the landscape and already delineated as covering the 72 miles from the Seneca River to Rome at the headwaters of the Mohawk, and an eastern segment with an increasingly steep downward slope over 126 miles from Rome to Albany on the Hudson.

Provision for financing was the next order of business.[16] The sections of the state to the north and to the south of the proposed route of the canal put up stubborn resistance to contributing anything, on the theory that only the areas through which the canal passed would receive any gain from it. Accordingly, the basic scheme, designed by George Tibbits, a member of the legislature, involved a combination of taxes aimed at those who would benefit most directly from the canal. Proposal is not disposal, and all taxes give rise to controversy and fierce attack from those whose pocketbook is about to be invaded. When the proposed legislation finally came on the floor of the legislature for consideration, these recommendations were no exception.

All real estate located within twenty-five miles of the canal was to be subject to tax, as earlier planning had provided. This recommendation stirred the greatest amount of opposition. Although enacted, the tax was never collected. As a contribution from the west, there was to be a tax on the sizable salt production in the area around Syracuse, which had been going on ever since the Iroquois had been the sole occupants of these lands. Many of the salt workers were Irish, who introduced both potatoes and corn soaked in brine into the local diet. These dishes are still a specialty in Syracuse.[17]

As the contribution from the east, another tax would apply to steamboat passengers, who would now be able to travel westward from New York and eastward from Lake Erie on water instead of on land. The commissioners also included a tax on an already existing if odd form of state revenue: the duties collected on imported goods sold at auction. Instead of the usual procedure, whereby merchants order foreign goods from abroad, since 1784 in New York State auctioneers received the imported merchandise and held public auctions to dispose of them, remitting the net proceeds to the foreign exporter. This procedure became increasingly active after the War of 1812, with individual auction houses doing as much sales volume as thirty or forty firms in normal trading activities. The commissioners also looked forward to grants of privately owned land from landowners eager to stay in high political favor.

The fundamentals of Tibbits's scheme showed unusual sophistication for the time. Tibbits admitted that the proceeds from all these taxes would produce no more than about $200,000, which "would allow of the canals [Erie and Champlain] progressing slowly." But if the tax revenues were perceived as collateral against ultimate repayment of funds borrowed to finance the canal, then matters "would allow [the canals] to progress rapidly, and leave a reasonable assumption that they would be completed in twelve years, estimating their costs at seven millions."[18]

In the end, the amount the state had to borrow exactly matched Tibbits's estimate, with no security other than this trickle of revenue from taxes, a promise of sufficient toll revenues to service these obligations, and the full faith and credit of the state itself if the tolls were to fall short. But construction moved ahead far faster than Tibbits anticipated, taking only seven years rather than twelve.

* * *

The selection of engineers to map out and design the construction of the canal was the most difficult task the commissioners had to confront, because there was no one in the entire United States who could properly style himself as a professional "engineer." The commissioners had originally turned to William Weston, the English engineer of the Western Inland Lock Navigation Company—including the handsome salary of $7000 a year—but Weston declined because of his age and ill health after wishing them well and predicting "the noble and stupendous plan . . . [would] have to fear no rivalry." He characterized the offer as the greatest honor he had ever received.[19]

Having no better choices to make, the commissioners named James Geddes and Benjamin Wright as engineers of the western and middle sections, respectively, and another engineer/surveyor named Charles Broadhead to the eastern section. Wright and Geddes were so good at their jobs that, when each made separate readings of one long stretch, coming from opposite directions, they came together with a space of only two inches between them.

Subsequently, they were joined by Canvass White, a twenty-seven-year-old wounded veteran of the War of 1812. Although White was an amateur engineer, his real passion was for research. He was convinced that the brief reports on the British canals by people like Elkanah Watson over the years were too superficial and that the canal engineers would need detailed information on the achievements of their English counterparts. In 1817, he would persuade Clinton and Wright to let him go to Great Britain, at his own expense, to examine how the English had solved the construction and design problems the commissioners were in the process of trying to resolve. The trip was an arduous one, not completed until 1818. White covered two thousand miles of the English canal network on foot, studying every detail that would be helpful for the work back home.[20] Once the canal was under way, White would produce one of the key technological feats of the whole enterprise.

Three of the commissioners and the engineers spent most of the summer of 1816 plodding one last time over the whole proposed territory, including the route of the Champlain Canal. As Clinton described this pilgrimage in a letter to a friend, "We have looked at all the difficult points, ascended mountains, penetrated forests, descended into wide-spreading and deeply excavated ravines, and have, on the whole, encountered more fatigue than I thought I would bear." But then he cheers up: "The result is most satisfactory. The work can be easily effected, and the utmost cost will not exceed our calculations."[21] In another account of the trip, Clinton described how the entire area, including the route to Lake Champlain, had been "explored, surveyed, and levelled . . . the routes of the canals had been actually laid out . . . perspicuous maps and profiles had been made . . . and the whole expense did not amount to 24,000 dollars."[22]* One of the most attractive features of the entire route was the frequency with which the canal would inter-

*The verb "level" refers here to a surveyor's measurements to find the heights of different points in a piece of land or an extensive area.

sect rivers and streams flowing northward toward the Great Lakes, which would provide an ample supply of water for the project all the way across New York State.

Clinton fails to mention the most remarkable feature of the entire effort: the outstanding performance of the novice engineers. Neither James Geddes nor Benjamin Wright was a professional surveyor. Neither had any education in engineering. Their only training had taken place in their long experience as judges, where they dealt on many occasions with surveys to settle disputes over properties. Before taking up his responsibilities on the canal, Geddes had used a leveling instrument only once, and then for just a few hours.

No wonder Commissioner Samuel Young was worried when Geddes began the complex task of surveying the summit near Rome where the Mohawk rises and begins its eastward journey toward Albany. "The whole project stands upon an attenuated thread [which] could cause our destruction," Young wrote to Clinton. "A Mistake in the summit level [would be] irrevocable damnation in this world."[23] Young need not have been so fretful. When the commissioners came to draft their report, they would proclaim their pride in these men, whose "versatile ingenuity [and] degree of care, skill, and precision in the delicate art of leveling . . . had perhaps never been exceeded."[24] Despite the lavish praise, the commissioners would continue to refer to Wright and Geddes as "surveyors" rather than as engineers.

* * *

On February 15, 1817, about a year after Van Buren's bill had been enacted, the commissioners presented the final report on their findings and decisions—all 174 printed pages of it, including accompanying documents with mile-by-mile details of the surveys and cost estimates. The report also included information gleaned from the technical literature in other countries as well as Samuel Young's imposing treatise on internal navigation.[25] The total expense of the Erie Canal was estimated at $4.9

million, covering a distance of 363 miles with eighty-three locks and eighteen aqueducts. The cost per mile worked out to $13,400, about $1600 less than Gallatin and Fulton had figured in 1808. The estimated outlay for the Champlain Canal was set at $900,000.

These were staggering sums for the world of 1817. The total outlays on Washington's Patowmack Canal, for example, had run around $300,000, while the Western Inland Lock Navigation Company had spent about $400,000. Seen from this perspective, the shock that sustained the opposition to the canal is readily understandable and the success of the campaign to build it is all the more remarkable.

Although Lake Erie lies 541 feet above the Hudson River, the ups and downs of the landscape along the way to Albany required locks to raise and then lower boats by a total of 661 feet. The commissioners also specified the proposed dimensions of the canal, which would be 40 feet wide on the water's surface and taper to 28 feet wide at the bottom, with a depth of 4 feet.

In the interest of provoking as little opposition as possible, the report accepted the conservative recommendation of the canal bill of 1816 that work at the outset should be limited to the middle section between Rome on the upper reaches of the Mohawk River and Montezuma on the Seneca River near Lake Cayuga, a distance of almost seventy-five miles. Here, where the landscape was relatively flat, the engineering would be simple and the cost modest at about $1.5 million. Referring to "counsels of prudence," the commissioners suggested bringing "the solidity of their opinions to the touchstone of experiment before the whole system is undertaken."[26]

But low cost and ease of construction were only part of what was involved here. By linking the western and eastern areas of New York State, this middle section of the canal would be the most strategically important component of the whole enterprise. The report predicted that "the greater part of the trade . . . from all these vast and fertile regions west of the Seneca lake, will be lost to the United States" unless this segment of the canal were to be completed.[27] Now New York would provide a far shorter, smoother, and less costly artery than the long and hazardous Canadian route to the Atlantic Ocean via Lake Ontario and the St. Lawrence River.*

When all was said and done, Commissioner Myron Holley wrote to his colleague Joseph Ellicott, "I cannot but believe that . . . the Legislature will be disposed to take efficient measures for the accomplishment of this great work."[28]

Holley was correct. A month later, a joint committee of the Senate and Assembly of New York State, acting "under the scoff and hiss of [the] general government,"[29] submitted to the legislature "an act respecting navigable communication between the great western and northern lakes and the Atlantic ocean." Nevertheless, all the old lines of opposition rose up in one last stand, and the debate was hot.

The New York City delegation was especially hostile, as usual, obsessed with anxieties about "cheap competition" from wares produced in the interior, concerned about paying taxes on which they would receive no conceivable benefit, and consumed with hatred for De Witt Clinton. They held fast even after Elisha Williams, a famously eloquent assemblyman from Columbia County, had reminded them that "if the canal is to be a shower of gold, it will fall upon New York; if a river of gold, it will flow into her lap."[30] In the final vote, the canal carried the day in the Assembly, 51–40.

The climax in the Senate was high drama. In the midst of the usual babble of debate, Martin Van Buren rose to speak. The political environment had undergone a sea change since he had sandbagged the legislation of 1816. His ally, Daniel Tompkins, was no longer on hand, having become vice president of the United States under James Monroe.

*The commissioners also provided detailed proposals for financing and for management of the canal's funds. For more detail, see Nathan Miller, *The Enterprise of a Free People*, pp. 65–71 and ch. 5.

De Witt Clinton just a few weeks earlier had defeated the redoubtable Peter Porter in the contest for his party's nomination for governor, in spite of Van Buren's vigorous efforts to stop him. Now Clinton was running for election on the Republican ticket—unopposed even by the Federalists. He would go on to win 97 percent of the vote, the remainder consisting of write-in votes for Tompkins.*

Before a hushed house, Van Buren startled his colleagues with a fervent appeal in favor of the canal, declaring that this vote was the most important he had ever given. All the old objections had vanished. After six years of commissions, explorations, surveys, and debates, the honor of the state demanded action. Indeed, Van Buren said, "our tables have groaned with the petitions of the people."[31] Clinton sat in the Senate chamber transfixed as Van Buren continued to develop his case for the canal. Van Buren himself later described the scene: "Altho' it was only two weeks since he had obtained the nomination for governor, which I had so zealously opposed and our personal intercourse was very reserved, he approached me when I took my seat, and expressed gratification in the strongest terms."[32] On April 15, the Senate approved the bill to create the Erie Canal, 18–9.

Van Buren's conversion to the canal had an ironic twist twelve years later, when he had attained Clinton's post as governor of New York. In a letter dated January 31, 1829, he urged President Andrew Jackson to "preserve the canals" against the onslaught of the "railroads," which he repeatedly refers to only in quotation marks. Ignoring the huge job creation capabilities of the railroads, Van Buren argues that the railroads are causing serious unemployment among "captains, cooks, drivers, hostlers, repairmen and lock tenders . . . not to mention farmers supplying hay for horses." Furthermore, the Erie Canal is essential for the

*One of Clinton's more interesting achievements as governor was to establish the second Thursday of November as a day of thanksgiving and prayer—a measure New York's first governor, John Jay, had been unable to accomplish because of claims he was trying to enlist the religious prejudices of the people in his favor.

defense of the United States. Worst of all "'railroad' carriages are pulled at the enormous speed of fifteen miles per hour by 'engines' which . . . [endanger] life and limb of passengers, roar and snort their way through the countryside, setting fire to crops, scaring livestock, and frightening our women and children. The Almighty certainly never intended that people should travel at such breakneck speed."[33]

* * *

One last hurdle remained before the canal could become a reality instead of just a shimmering vision: the new legislation had to be approved by the Council of Revision. This curious creation of the constitution of New York State had the power to veto any legislation its members deemed inappropriate; its decisions could be reversed only by two-thirds majorities in the Assembly and the Senate. The council consisted of the governor, the chancellor—a specialized judgeship—and the three justices of the State Supreme Court. New York also had a Council of Appointment to approve or disapprove all the governor's nominees for appointive office, which consisted of four senators, one from each quarter of the state, elected by the Assembly every year. The governor could vote to break a tie among the four members but his appointments were otherwise at their mercy. The constitution of New York State had created these two bodies as a means of limiting the power of the executive and the legislature, which indeed they did.

On this occasion, the Council of Revision consisted of one confirmed enemy of the canal, John Taylor, who was serving out the remainder of Tompkins's term, and two equally determined supporters, our old friend Justice Jonas Platt and his Supreme Court colleague Joseph Yates. The other two council members were Chief Justice Smith Thompson and Chancellor James Kent.

Kent and Thompson made an interesting combination. Kent was then fifty-four years old. He had graduated from Yale at eighteen as a member of Phi Beta Kappa in a class that included several future members of Congress, two United States senators, a minister to Europe, chief justices of Vermont and Connecticut, and three governors of Connecticut. After teaching as the first professor of law at Columbia, Kent was a judge and then became chancellor of New York State, where he dealt primarily with matters of equity, a system of law that differed in its structure from the English common law.* He was such a distinguished member of the legal profession that the law school at the Illinois Institute of Technology bears his name and his portrait hangs today over the judge's bench at the Court of Appeals of the State of New York—now the state's highest court.[34]

Thompson and Kent had known each other for more than twenty-five years. Thompson, a Princeton man born in 1768, began his legal education at the firm run by Kent while he was still in private practice. Nevertheless, there was evidently no love lost between the two men, as Kent described Thompson as "a plain, modest, humble, ignorant young man with narrow views and anti-federal politics."[35] But Thompson went on to serve as a state legislator before being named to the State Supreme Court in 1802.

As the Council of Revision's discussion opened, Thompson expressed his opposition to the canal bill, claiming that it gave the commissioners arbitrary powers over private rights. Kent was also opposed, but for a different reason. He was concerned about committing the state to this huge project before public opinion was more clearly and more emphatically in favor.

After each member had expressed his views, more extended and active debate began, leading to what Platt described as "a more temperate and deliberate examination of the bill" that overcame some of the objections of the chancellor and the chief justice. The bill's future had just begun to look more promising, when who should walk into the

*Equity, even in its more limited modern sense, is still distinguished by the principle that no wrong should be without an adequate remedy.

meeting room but Vice President Daniel Tompkins, claiming he happened to be in the neighborhood and thought he would drop by.

Coincidence or not, Tompkins was fully prepared to participate in the proceedings and lost no time in coming out against the canal bill. He contended, based on his knowledge of the Washington scene, that only a truce had been established with Great Britain. Instead of wasting the credit and resources of the state on a project of such uncertain success, he argued, the legislature should be devoting all of the state's resources toward preparing for war. "My word for it," he averred, "we shall have another war with [England], within two years."

Tompkins touched a sore spot with Chancellor Kent. According to Platt, the chancellor rose from his seat and declared with great animation, "If we must have war, or have a canal, I am in favor of the canal, and I vote for this bill." Kent gave the pro-canal group the majority and so the canal bill of 1817 became law.

* * *

Construction began at sunrise on July 4, 1817, just outside of Rome, where a large crowd had gathered along an unremarkable stretch of grass and trees. They had come to listen to the usual litany of speeches, but especially to observe Judge John Richardson, the canal's first signed contractor, stationed behind a team of oxen with his hand on his plow. A town with an imperial name was about to witness the birth of a project that would turn New York into an imperial state.

Commissioner Samuel Young began the proceedings with a speech remarkably short for such an auspicious occasion. Young prophesied, "By this great highway, unborn millions will . . . hold a useful and profitable intercourse with all the maritime nations of the earth."[36] Then the cannon at the nearby Rome arsenal boomed and Judge Richardson's oxen moved forward, driving his plow as it made the first cut in the earth to inaugurate the construction of the Erie Canal. Almost ten years had passed since Jesse Hawley predicted, "If the project be but a feasible one, no situation on the globe offers such extensive and numerous advantages to inland navigation by a canal, as this!"

The project had finally been proven feasible. Three days earlier, on July 1, De Witt Clinton had taken the oath of office as governor. Clinton's Ditch was about to be transformed into the Great Western Canal—and "both ways to oncet," as Yorkers liked to describe it.[37]

PART III

The Creation

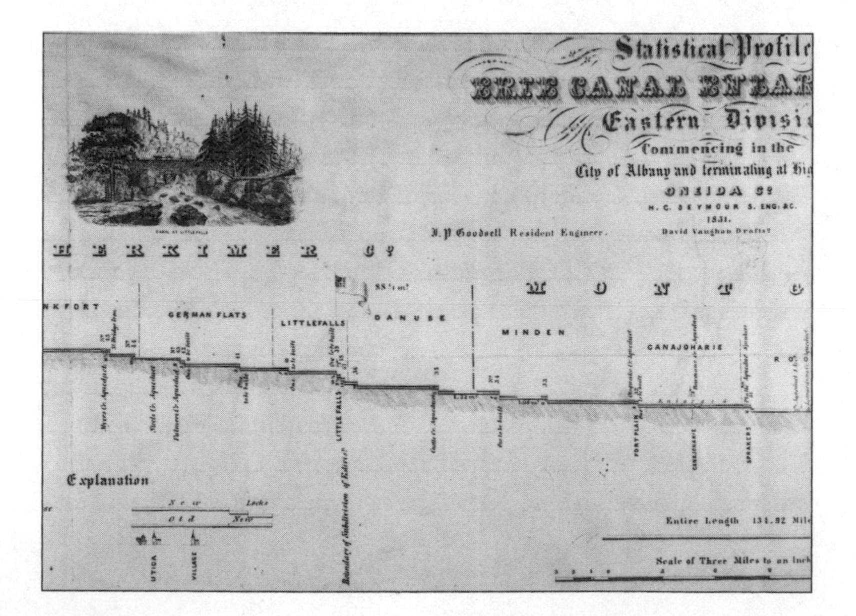

CHAPTER 11

DIGGING THE DITCH

In the summer of 1831, Alexis de Tocqueville, the famous French observer of the American scene, traveled by horseback across the state of New York, generally following the route of the Erie Canal from Albany to Buffalo. As he approached Albany, he was appalled at the countryside, so dramatically different from Europe: "Man still making clearly ineffective efforts to master the forest," he noted in his diary. "Tilled fields [with] trunks in the middle of the corn. Nature vigorous and savage."[1] And this was six years after the Erie Canal had been completed and fourteen years after the ceremony at Rome that launched it.

As soon as he ventured even a short distance away from the developed areas along the path of the canal, de Tocqueville found only the frightening wilderness. Here is how he described the countryside as he rode fifteen miles northward from the canal at Syracuse one July afternoon into "one of those deep forests of the New World whose sombre savage majesty strikes at the imagination and fills the soul with a sort of religious terror":

> How can one paint such a sight? . . . Nature has sown pell-mell in incredible profusion the seeds of almost all the plants that creep over the ground or climb above the soil. Over our heads

> stretched a vast dome of vegetation. Below this thick veil . . . there was one vast confusion: a sort of chaos. . . . Generations of trees have succeeded one another there through uninterrupted centuries and the ground is covered with debris. . . . It is like a fight between death and life. Sometimes we happened to come on an immense tree that the wind had torn up by the roots, but the ranks are so crowded in the forest that often despite its weight it had not been able to make its way right down to the ground. . . . A solemn silence reigned in the midst of this solitude. . . . Man was missing from the scene.[2]

Digging a ditch four feet deep and forty feet wide with hand labor through hundreds of miles of this primeval forest was the greatest challenge the builders of the Erie Canal would have to confront. The only sources of power were the traditional ones—human muscle plus horses or oxen. There was just one innovation of importance that had become available only recently: the newfangled blasting powder manufactured by a young Delaware chemical company, E. I. Du Pont de Nemours, providing more bang for the buck than the black powder used over the centuries but nowhere near as powerful as the dynamite and nitroglycerine that would be developed in the future.

There was no precedent for such an undertaking. The English, French, and Dutch canals had been cut through cleared and populated countrysides spotted with only small and randomly distributed forests. Although the engineers could draw upon the ample experience of earlier canal builders when confronted with crossing the rivers astride the canal's path or constructing the locks, nothing in the bulging portfolio of notes that Canvass White would bring back from England in 1818 contained even a hint of how to conquer the forest. The Erie Canal's projected landscape was so untamed when construction began that there was no scheduled stagecoach line west of the headwaters of the Mohawk at Rome, which meant that the stages traveled only 125 miles westward from Albany before turning back toward the Hudson. Fewer than ten towns punctuated the middle section between Rome and the Seneca River to the west, where the first wave of construction was planned.[3]

At the proposed terminus of the canal at Lake Erie, the neighboring towns of Buffalo and Black Rock contained only 700 people between them. Indeed, in the 14,000 square miles of western New York, bounded by the Seneca River on the east, there were fewer than eight people per square mile in 1820, three years after work on the Erie Canal began.[4] Everything else was "sombre savage majesty," much as it would appear to de Tocqueville in 1831. The single significant break in the forest was at the western end of the middle section—and there only because of the insect-laden muck across swamps and marshes.

* * *

De Witt Clinton was now the newly elected governor of New York and still a member of the prestigious canal commission. He was in full control of the great adventure. The record would turn out to be even more impressive. Taking his second and third terms as governor into consideration, De Witt and his uncle George Clinton between them sat as governor of New York State for thirty of the fifty-two years between the American Revolution and De Witt Clinton's death in 1828. By an odd coincidence, these two Clintons had been preceded by another George Clinton, an English cousin, who served as the royal governor of New York from 1743 to 1753. The American George, in addition to his long stint as governor of New York, was vice president during Thomas Jefferson's second term and again, under James Madison. He died in office in 1812.

Although he was the dominant member of the canal commission, De Witt Clinton set to work with an impressive group of men who appear to have had a remarkably clear sense of how to go about their unique and imposing responsibilities. Their cooperation in these tasks in the midst of the political combat constantly swirling around them makes their achievements even more impressive.

In order to keep themselves beyond suspicion and to work as effectively as possible, the commissioners agreed to limit their functions to the actual construction of the canal. They transferred the financial responsibilities they had carried up to this time, such as managing the various taxes set up to finance the canal and borrowing funds, to a new body named the Commissioners of the Canal Fund, consisting of the state comptroller, the treasurer, the secretary of state, the attorney general, the surveyor general, and the lieutenant governor.

When it came to everything else, any large modern corporation could envy the commissioners' style of management and organization. Considering the uncongenial landscape to be crossed, the sheer magnitude of the whole project, and the lack of experience in designing, organizing, and supervising anything of this nature, their performance was remarkable. These five men—De Witt Clinton, Stephen Van Rensselaer, Joseph Ellicott, Myron Holley, and Samuel Young—managed their priorities, encouraged the introduction of totally new techniques for overcoming the formidable obstacles of the forests, and met their deadlines from the beginning. Clinton boasted how effectively they warded off "frivolous calls for information [such as] a statement of their expenses . . . proposed by . . . a gentleman [who] had not the magnanimity to overlook a disappointment in not obtaining the office of their secretary."[5] Holley and Young, the designated full-time commissioners, received salaries of $2500 at first, but their pay was subsequently reduced to $2000 so that they would not appear to be using their positions to enrich themselves; these two men were also serving as members of the legislature at the same time.[6]

By the end of 1817, only six months after the first earth was turned, the commissioners had some fifty contracts signed with the total expenses well within projected estimates, fifty-eight miles of work under way, a thousand men on the line, one job completed and fully paid, and about fifteen miles of canal fully operable. The achievement, as Clinton informed Joseph Ellicott, had exceeded his "most sanguine expectations."[7]

The commissioners' critical decision was to assign as much of the digging as possible to local people instead of enlisting a limited group of large-scale contractors. The locals were most familiar with their countryside, knew best how and where to recruit a labor force, and would take greatest pride in building the stretch of canal near their homes. In addition, by letting out a large number of small contracts covering only short distances, the commissioners were able to diversify the unavoidable risks of using contractors who had little or no experience at canal building but who brought the knowledge of a wide range of occupations, with varied skills and knowledge of the work to be done.

Most of the stretches of canal assigned by these individual contracts were bounded by brooks or ravines so that each contractor would in effect be constructing a little canal of his own, quite separate from the canals his neighbors on either side would be building. The contracts also provided that no payments could be made until the entire job was completed and had passed inspection by one of the engineers, but the commissioners permitted informal advances for out-of-pocket expenditures for stores of food and materials as well as monthly payments to cover ongoing labor costs.

The going wage for labor was $12 a month, although the men were often provided with food (and occasionally with drink as a reward when goals were attained) and crude sleeping quarters as well (by way of comparison, members of the legislature then received $3 a day plus another $1.50 for each day's travel).[8] When pay included free travel and overnight accommodations, the expression was "fifty cents a day and found." According to a story in the *Albany Gazette*, on paydays Commissioner Myron Holley in his carriage would ride up and down in the section for which he was responsible, doling out the necessary amounts to each of the local contractors, who could then pay their crews.[9] In the beginning, as the commissioners reported in 1819, most of the workers were "born among us," including a few freed slaves.[10] A large contingent of Irishmen has become a glowing part of the canal's folklore, but they did not come upon the scene until later.

* * *

The first step in building the Erie Canal took place when the crews defined the area to be cleared of the forest by driving red stakes sixty feet apart along the whole line of the proposed canal and then another set of stakes precisely forty feet apart to mark off the area to be excavated for the canal itself. The difference left room for the towpath and the berm (the side of the canal opposite the towpath).

The basic design of the canal recognized that it would be too shallow and probably too narrow for the kinds of steamboats then in operation, whose side wheels would churn up the muddy bottom only four feet below the surface while the wake would damage the canal walls from the constant waves banging against them. Sail power would be inappropriate because winds in one direction would not be suitable for boats moving against the wind and the canal would be much too narrow to leave any room for tacking. Poling would be too slow and impossible for heavy loads.

The only practical form of motive power was horses and mules to tow the boats along behind them, the same method the English had adopted for their canals. The need for a towpath along the entire length of the canal explains why the total width was designed to be twenty feet wider than the ditch itself. The towpath was often narrower than twenty feet, but the berm also had to be provided for to the extent possible so that people could move on the other side as well.

As originally planned, the middle section would be the easiest to build. Part of it, the so-called Long Level along the relatively flat country from Utica to Syracuse, was a stretch of fifty-nine miles, designed to be built without a single lock. This stretch was the longest canal level in the world without locks. Early progress was everything anyone had hoped for. When the first snowfalls forced work to halt at the end of 1817, the commissioners could report, "Much useful experience has been acquired in the course of the season."

The formal language obscured the drama of striking inventions and technological innovations the crews on the job developed to meet problems whose magnitude no one had ever encountered before. The most interesting achievements involved what the commissioners described, again with notable understatement, as "many valuable improvements . . . made in the method of grubbing standing timber." Grubbing—digging up or digging out de Tocqueville's "vast confusion [of] generations of trees"—was the most important element of the process. Not a yard of ditch could be dug until the workers had cut down the countless thousands of trees, chopped them up into movable sizes, uprooted the stumps, and then carted away the staggering mess of logs, branches, and leaves.

Neither scientists nor engineers devised the ingenious laborsaving solutions to these tasks. Rather, the inventors were the men doing the hard work on the job. They remain anonymous. When the great celebrations at the completion of the canal took place, these men would continue to be unheralded. Yet the scale of their achievements was not to be denied.

Most of these remarkable laborsaving innovations on the canal owed their success to the principles of Archimedes, the great inventor and mathematician of ancient Greece, who has received the credit for inventing the lever—a device based on the theory that a small amount of force moving a relatively large distance could translate into a greater amount of power delivered over a smaller distance. Most levers are boards set on a fulcrum nearer one end than the other—a kind of unbalanced seesaw—or are gears where a small wheel turns faster than a larger wheel to which it is connected.

One such clever invention on the Erie Canal dramatically reduced the time-consuming labor needed to chop down the massive trees in the forests, which would often have involved more than a hundred swings of an axe before the job was done. This gadget consisted of a wheel wound around with a cable and mounted so it could spin freely. When the loose end of the cable was attached to the top of a tree, one man all by himself could fell it by turning the wheel until the cable bent the tree over so far it finally broke free of its stump and crashed to the ground.

But the stump was still there, its thick roots reaching in all directions under the surface of the earth. The contraption contrived to uproot the stump was a truly formidable device. The basic machinery consisted of two wheels, each sixteen feet in diameter, set on either side of a huge axle thirty feet long and twenty inches in diameter. Another wheel, with a diameter of fourteen feet, was mounted in the middle of the axle. As the diameter of this wheel was two feet smaller than the outside wheels, it did not touch the ground and could spin freely. In order to drag a stump and its roots completely out of the earth, the men first braced the two outside wheels of the axle with heavy rocks so that the device would be stationary. Then they wound a chain around the axle and fastened the chain to the stump. After wrapping one end of a strong rope several times around the fourteen-foot wheel in the center of the axle, the men attached the other end of the rope to a team of horses or oxen. As the animals moved forward, pulling the rope behind them and turning the wheel, the tremendous pressure on the chain yanked the stump and its roots free of the earth. According to one authority, one of these contrivances with a team of four horses and seven laborers could grub thirty to forty large stumps a day.[11]

After all this, the ground was ready for the excavation to begin, but no one had built anything as enormous as the Erie Canal before. In the early stages, the men followed the so-called European method of digging with spades and carting the excess earth away on wheelbarrows, an approach hallowed by thousands of years of use. Pressed to find more efficient ways to dig and remove the excavated earth, the men discovered they could save a lot of time and effort by replacing spades with horse-drawn plows. These plows carried heavy and sharpened pieces of iron that sliced through the roots as they were pulled along, tearing apart the earth. In addition, the teams of horses constantly carrying away heavy loads of excavated earth were packing down the ground along the side of the canal into solid banks much less vulnerable to leakage than with the traditional method.

All of this engineering improvisation by untrained talents would one day be known as Yankee ingenuity. But these innovations had an awesome quality as well, as signs of the growing force of man over nature in the budding Industrial Revolution. Cadwallader D. Colden, the grandson of the first man to dream of an Erie Canal nearly a hundred years earlier, put it this way in 1825: "Indeed, to see a forest tree, which had withstood the elements till it attained maturity, torn up by its roots, and bending itself to the earth, in obedience to the command of man, is a spectacle that must awaken feelings of gratitude to that Being, who has bestowed on his creatures so much power and wisdom."[12]

* * *

De Witt Clinton was eager to use the impressive achievements of the first season as a base for pressing forward aggressively during the year ahead. Promising that "the completion of these stupendous works [would] spread the blessings of plenty and opulence [to] the most distant parts of the Union and command the approbation of the civilized world," he insisted that "we are required to persevere."[13] As he saw it, there was not a moment to lose: he had his heart set on the ambitious target of a complete and fully operational canal by the end of 1824. It was no coincidence that 1824 would also be James Monroe's last full year as president (on the likely assumption he won reelection in 1820), bringing to an end the Virginia dynasty that had accounted for four of the first five presidents of the United States. The presidential contest in 1824 was wide open.

Clinton's secretary and friend, Charles Glidden Haines, provided strong support for Clinton's promises in his extended pamphlet of October 1818. Early in this document, Haines echoes George Washington's concerns as the prime motivation for building the Erie Canal: "Our mountains must be politically annihilated. Our sectional barriers must be swept away by a moral arm, whose power is resistless. . . . Nothing but this, can perpetuate that union which is to guarantee our future national greatness. Nothing but this, can preserve those popular institutions which are sealed with our fathers' blood."[14]

The canal commissioners responded. Paperwork moved ahead smoothly in 1818. By the late summer, contracts had been signed for the entire length of the canal except at a few spots around projected structures like aqueducts or bridges. In anticipation of the great day when the waterway would be approaching completion, Clinton arranged for a survey and drawing of plans for a harbor at the mouth of Buffalo Creek on the shores of Lake Erie.

At the construction site, however, the hard work on the canal did not move ahead smoothly, at least not at first. The winter snows had been unusually heavy, which meant a late melt. The melting had barely begun when the heavens dumped so much rain over New York that the crews could not resume full operations until late May. All the work had to wait until the mud and puddles dried out. But after that, the advance moved along at such a vigorous pace that most of the lost time was made up, with nearly four thousand men and fifteen hundred horses engaged by the time the digging season came to an end with the return of winter weather.[15] Progress was so convincing the commissioners were willing to project completion of the whole middle section by the end of the next digging season in 1819.

Although the clearing of the land and the excavation of the canal were now making satisfactory headway, the canal staff had not yet found a totally waterproof material, good for all seasons, to seal the spaces between the neatly arranged stones lining the sides of the excavation, the locks, the culverts, and the aqueducts. The available common quicklime in use was unstable, breaking down under any sustained pressure. As a consequence, leaks as well as rot were already making their appearance in the canal walls and the sides of the locks, requiring constant relining of the surfaces. Without a replacement for the quicklime, the canal could conceivably end up as nothing more than a pile of mush. Clinton, ever the voracious reader, had come upon a Roman cement for this purpose in *The Repository of the Arts* and obtained some samples from Italy. This material appeared to do the trick, but importing massive amounts of it across the Atlantic Ocean would have been prohibitively expensive.

Then the young engineer Canvass White heard that the contractors responsible for building culverts and aqueducts on the stretch between Syracuse and Rome might have come upon an answer to the puzzle in the vicinity of Chittenango, a little village about twenty miles east of Syracuse.* They were using local limestone that would not "slack"—that is, it would not disintegrate when wet—unlike all other limestone tried so far. White and Benjamin Wright traveled to Chittenango as soon as they received this news and set up a meeting with a local scientist named Andrew Barto in order to perform a full experiment with the material.

The encounter took place at Elisha Carey's barroom in the middle of town, an unlikely locale for a scientific experiment but a good place for the impatient spectators to calm their nerves.† Dr. Barto arrived with a handful of moist mortar made from the limestone, which he now mixed with sand, rolled into a ball, and deposited into a pail of water, where it would sit overnight. When the group gathered at Elisha Carey's the next morning to examine the ball of cement, they found they could roll the ball across the room. It was hard and solid as a rock. Dr. Barto declared this cement was equal to anything they could find in Italy or Holland.

White conducted more experiments with the Chittenango cement until he was satisfied he had the perfect mixture to make it waterproof under all conditions. Then he saw to it that a factory was set up just outside of Chittenango to grind and manufacture the material, at a price to the commissioners of $3.50 for a barrel of five bushels of cement.

This was a development of the greatest importance. In addition to

*Chittenango is the birthplace of L. Frank Baum, author of *The Wizard of Oz*. A yellow brick road runs through the center of town.
†In a nice bit of irony, the site of Elisha Carey's tavern is now an engineering school.

the 500,000 bushels of cement used in building the Erie Canal, the cement was exported in large quantities over the years to other states with similar needs. But the story has an unhappy ending for Canvass White. Although the canal commissioners promised White a "just" compensation, they procrastinated for a long time while keeping White's hopes up with an appropriation—but not a payment—of $10,000. It was all talk and no action. White ended up empty-handed.[16]

Despite this shabby treatment by the authorities, White continued to make significant contributions to the success of the canal. He prepared the plans for all the mechanical structures such as locks and gates, as well as some of the canal boats, even while continuing to provide expert surveying work. After all this, the most he ever got for his trouble was an accolade in the commissioners' report of March 12, 1821, in which they declared that White's usefulness, "from the beginning, has been constantly increasing with the progress of our labors, by his continued assuidity and increasing knowledge."[17]

* * *

The following year, 1819, looked at first as though nature would be a troublemaker once again. The winter months were usually the easiest time to bring material and supplies up to the canal sites, because sleighs running smoothly on snow were a lot more efficient than wagons rattling along on rutty roads. But not in the winter of 1818–1819. At first the snowfall was too meager to permit sleighing. When the snows finally made their appearance, they arrived in what seemed to be one endless blizzard, blocking the sleighs until March. Although matters improved with an unusually dry spring, which was a big help, the extended summer heat and humidity were brutal and the mosquitoes a scourge, especially in the marshy country between Syracuse and the western end of the middle section at the Seneca River. The work crews had joked at first about the easy digging through the soft earth of the marshes, but the good cheer vanished when clouds of voracious mosquitoes descended on them. At least a thousand men came down with fevers, no doubt including a high incidence of malaria, with medical care amounting to little more than snakeroot, green pigweed, or bleeding. Some jobs shut down completely until the cooler weather finally chased the mosquitoes away.*

The mosquitoes must have been a terrible curse everywhere in those years. In 1818, De Witt Clinton and his wife, Maria, who was already in poor health, planned to spend the summer months on Staten Island. The insects were merciless, and Maria fell gravely ill. In search of relief, the family moved up to Mt. Vernon in Westchester County but encountered swarms of mosquitoes there, too. Maria Clinton passed away at Mt. Vernon on July 30 at the age of forty-two, after twenty-two years of marriage.

Never one to let events overwhelm him, Clinton married for a second time nine months later. In between Maria's death and his remarriage, he took a terrible fall, which left him with a bad limp and limited physical activity for the rest of his life, but he maintained the frequency of his ceaseless travels of inspection by horse and by stage up and down the rough country along the path of the Erie Canal.

* * *

Meanwhile, the legislature was in an unusually cooperative mood in response to the commissioners' reports of the introduction of the ingenious laborsaving inventions and White's miraculous cement. In delivering the governor's opening address to the session of 1819, Clinton declared that the work already accomplished had demonstrated the feasibility of the canal and the planned revenue would without doubt cover the "necessary expenditure." Best of all, the effort promised huge payoffs:

*When construction by the United States began on the Panama Canal in 1904, the mosquito attacks were so violent and yellow fever and malaria so prevalent that the Americans employed over a thousand men just to cut grass and clear brush in order to make the working environment as inhospitable as possible to mosquitoes. See David McCullough, *The Path Between the Seas*, p. 573.

the movement of a ton of commodities costing a hundred dollars a ton to move by land "will not exceed ten dollars a ton . . . when the great western canal is finished."[18]

The authorities took the big step of authorizing completion of the entire canal on April 7, 1819. This meant that work in the 1820 season would begin moving outward in both directions from the ends of the middle section, westward from the Seneca River to Lake Erie and eastward from Utica down to Albany on the Hudson. There was more good news. The state's ample inflow of funds from borrowings allowed the legislature to suspend the tax on lands adjacent to the canal; this tax had been enacted but never collected in the original legislation in 1816. The legislature also provided that men working on the canal would be exempt from duty with the militia (frequent call-ups had been so disruptive to the labor supply that scheduling and sustaining canal work was often impossible). Plans for the harbor at the mouth of Buffalo Creek had advanced to a point where the legislators approved an appropriation of $12,000 for work to begin.

With planning for the entire canal now official, the authorities could finally bring to a conclusion the rocky road of negotiations for a grant of land to the state by the Holland Land Company, the owner and proprietor of by far the largest amount of acreage in the western part of the state. In 1810, even before the legislature had appointed the first canal board, Agent General Paul Busti of the Holland Land Company had offered as a gesture of goodwill "every other Lot adjoining each side of the said Canal in its passage through the Holland Land Company's lands."[19] As the lots were each 160 acres, the offer came to about 18,000 acres that the state would be free to sell to help finance the canal. Four conditions were attached: the canal would have to extend all the way to Lake Erie, it would have to be big enough for boats carrying up to twenty tons of freight, the canal would have to be built with public money only, federal or state, and no land would be released until construction had actually started. The state considered this offer for two years before rejecting it, maintaining that Busti's conditions were too constraining.

Thomas Eddy then pressed Ellicott, in his role as resident agent of the Holland Land Company, to renew negotiations by offering a larger grant than in the original proposition. Ellicott responded with what appeared to be a generous new offer, this time for 100,632 acres, but located south of the proposed route. The company stipulated that the lands were to be valued at a dollar an acre and the total value offset against any taxes the company would incur for financing the canal. And then they added the condition that they must be able to reclaim the land if the state had not completed the canal by the end of 1827. The offer may not have been anywhere near as generous as it looked, quite aside from the conditions. Ellicott referred to the offer in a private communication as "100,000 acres of mountains [but the greatly increased size of the offer] *seems* to evince greater liberality."[20]

The authorities in Albany nevertheless accepted this offer in November 1813. Then the deal fell apart the next year when the legislature canceled the authority of the commissioners to borrow $5 million from Europe, causing Clinton to complain, "The commissioners were thus frittered down into a board of consideration . . . without power and without money." Undaunted, Busti renewed his proposition when the work finally got under way in 1817, even though, as he confided to Joseph Ellicott, "the more I consider the nature of the undertaking and compare it with the temper of our State & general government the more I grow incredulous of its ever be[ing] perfected if begun."[21]

This time the transaction went through. Twelve years later, when the state finally got around to selling this huge tract, the proceeds came to a piddling $28,000, or 28¢ per acre.[22]

* * *

By late summer 1819, the middle section was about 95 percent complete. The great moment had arrived to let the waters into the canal. On October 22, the sluices were opened to allow water to fill the ditch over the sixteen miles from Rome to Utica. Elkanah Watson was impressed with the rate of progress, commenting that this date was two years, three months, and eighteen days from the Fourth of July 1817, when work had commenced. Now, according to Watson, "The scene was truly sublime. . . . It was impossible for stupidity itself, not to have been electrified on this joyous occasion, and to stretch their opaque minds from Erie to the ocean."[23]

In a letter to the *Albany Gazette*, a local citizen described the scene from a bridge across the canal near its eastern extremity at Utica as "a sight that could not but exhilarate and elevate the mind. . . .The waters were rushing in from the westward and coming down their untried channel toward the sea. . . . As this new internal river rolled its first waves through the state . . . the people were running across the fields, climbing on trees and fences, and crowding the bank of the canal to gaze upon the welcome sight. . . . It was dark before [the waters] reached the eastern extremity."[24]

A special boat named the *Chief Engineer* in honor of Benjamin Wright had been prepared for the occasion. The *Chief Engineer* was sixty-one feet long, seven and a half feet wide, and light enough to draw only fourteen inches of water, which meant it could be pulled along by one horse attached to a tow rope of about sixty feet in length. The *Gazette*'s informant was struck by the contrast between this modern means of transportation and the labored trips for boats of this size in the old days on the Mohawk River. He went on to describe "this new Argo, which floated triumphantly along the Hellespont of the west, accompanied by the shouts of the peasantry and having on her deck a military band." The new Argo made the sixteen-mile trip from Rome to Utica in four hours, at the speed of four miles per hour that would become standard throughout the entire length of the canal.

The following morning in Utica, as "the bells began a merry peal," the commissioners and about fifty assorted celebrities climbed into their carriages at Bagg's Hotel. As the drums rolled and the huge crowd of spectators roared their welcome, the commissioners reached the canal at exactly 9:15 a.m. and boarded the *Chief Engineer* for the trip back up to

Rome, where they arrived four hours later to enjoy the enthusiasm of more cheering crowds.

* * *

So much for the good news. While the drums were rolling and the crowds were shouting at the Erie Canal, the rest of the country was in deep trouble. What had started off as a modest setback in economic activity in 1818 suddenly turned into a full-fledged crisis—"panic" was the word used in those days—marked by soaring interest rates, widespread business distress, rising unemployment, cascading commodity prices, a collapse in real estate values, and an epidemic of bank failures. This shocking sequence of events would repeat itself many times in the future, on occasion with remarkable fidelity, but the crash of 1819 was the first home-grown depression the nation had experienced and therefore all the more frightening for those who experienced it.

The canal was a shining light in this tidal wave of darkness. Without the stream of dollars and the jobs provided by New York's great venture, the economic setback would have been even worse than it was. More than a century later, economists liked to refer to this cushioning effect as Keynesian deficit spending or expansionary fiscal policy. To the citizens of New York State in 1819, the canal was a plain everyday blessing.

But the blessings worked in both directions. The disasters sweeping across the rest of the nation turned out to be a stroke of luck for the Erie Canal. The story is worth telling in detail.

CHAPTER 12

BOOM, BUST, BONDS

Eighteen seventeen had been a remarkable year. Construction on the Erie Canal got under way after more than a decade of resistance and struggle. An American shipping company, the Black Ball Line, announced on October 24 that it would be running transatlantic ocean liners on a regular, announced schedule from Liverpool, the gateway to the British Midlands, and had selected New York City as their American port of choice. The New York Stock Exchange had a new constitution to replace the famous 1792 agreement arrived at by twenty-four brokers under Wall Street's buttonwood tree. Stock prices soared by 20 percent as the brokers and traders moved indoors from the street curbs for the first time. The birth of American investment banking followed soon after. In the economy at large, two years of vicious postwar deflation in the prices of goods and services came to an end, bringing welcome relief to people who owed money and to young business firms struggling to make a profit.

It was no coincidence that the roaring stock market and the ground breaking of the Erie Canal should have taken place in the same year. With the war behind them and growth and development occurring in more and more places in the land, Americans were in a buoyant mood about their future. A venture that had been perceived as highly risky just a short time earlier now appeared as something that could not fail. But we cannot grasp the full significance of the interaction between the canal and the exuberant boom without a brief history of the boom and its aftermath.

* * *

The business boom brought a wave of speculation in its wake, affecting many different areas of the nation. In both the cities and the countryside, prices of land and real estate were climbing right along with stock prices.* The amount of debt on land private citizens had bought from the government increased from $3 million in 1815 to nearly $18 million by 1819.[1] Federal construction outlays leapt from less than $1 million in 1816 to over $14 million two years later.[2] There was an accompanying burst of growth in turnpikes, or toll roads built, financed, and managed by private corporations. Imports surged ahead of exports, financed by shipments of specie abroad and by borrowing from foreign merchants.

How could such an ebullient economy have emerged out of the disruptions and disturbances of the War of 1812 and the deflationary pressures of the early postwar years? In an ironic twist of history, the story of the boom begins at one of the darkest moments of American history, August 24, 1814, the day the British captured the capital and set Washington on fire, leaving the president's mansion a blackened wreck.

President Madison had left the mansion on horseback early in the

*Here are some guidelines to provide perspective for the figures throughout this chapter. In terms of purchasing power, $1 in 1815 bought the equivalent of about $12 today. Nominal gross domestic product (GDP, or total national production of goods and services) in 1815 was $700 million; today it is over $11 trillion. GDP per capita today is more than thirty times greater than in 1815. The War of 1812 consumed about 13 percent of GDP over three years. On a per capita basis, the Revolution had been four times as costly. At this writing, the war in Iraq is taking less than 1 percent of GDP. Estimates (except for Iraq) from William D. Nordhaus, *The Economic Consequences of a War in Iraq.*

morning of that very hot August day, well aware the British were approaching Washington and also aiming for Baltimore. He headed first for the navy yard to assure himself about conditions at Chesapeake Bay, after which he planned to ride into Maryland to consult with the general in command of the nearby battlefield at Bladensburg, six miles beyond the navy yard. By the time Madison returned home about three o'clock in the afternoon, he had spent more than eight hours in the saddle, probably without anything to eat since early morning. But there was to be no rest for him.

During his absence, a small contingent of British had moved on Washington from the outskirts and set to work burning the city. In a performance whose elegance any professional terrorist might envy, the British restricted their arson to public buildings while sparing private property.* Rumors were soon circulating that such meticulous choice of targets must have been the work of a traitor who had identified the public buildings in advance to the British commander, General Robert Ross. Ross's superior, Lord Bathurst, took a different view of this performance and warned Ross that at Baltimore, the next objective, he should "make its inhabitants feel a little more the effects of your visit than what has been experienced at Washington."[3]

Madison found his home a shambles. Worse, no one could tell him where his wife, Dolley, had gone. She had escaped early on in her favorite carriage, but not before rolling up Gilbert Stuart's full-length portrait of George Washington and entrusting it to the care of employees in the house. She spent the rest of the day and the night at the home of friends in Virginia, where she had as little idea of what had happened to the president as he knew of her fate. He ended up staying with a different set of Virginia friends. The president and the first lady did manage to reunite the following morning.

*Luckily, employees at the State Department had already fled with a mass of important documents, including the Declaration of Independence and the full record of deliberations at the Constitutional Convention.

After assuring himself of Dolley's safety, Madison set off once more to confer with his general on the road to the defense of Baltimore. Sixty-three years old, and in less than robust health, he would spend four more days and most of the nights in the saddle before returning to Washington, and then he had to struggle to restore a government out of the chaos the British had left behind.

Panic broke out everywhere as the terrible news spread across the nation over the days that followed. As so often happens in panics, few things can relieve high anxiety more effectively than hard cash. People rushed to the banks clutching the paper notes that constituted most of the money supply in those days, demanding specie—coins of silver or gold, or real money you could bite into—in place of their paper banknotes. Although the Constitution had given Congress the power to regulate the currency of the United States, there was no national currency, nor would there be one until the Civil War. The only money in the system was the lavishly printed paper notes issued by individual banks as the proceeds of loans or to redeem customers' deposits, and a modest amount of minor coins that served as small change.

Anyone holding a banknote had the right to demand specie in exchange, but the limited supply of specie soon vanished as gathering hysteria in the wake of the disaster in Washington provoked more and more frightened citizens to rush to the bank to cash in their banknotes. Only the lucky few who managed to be at the head of the lines to the bank windows walked away with those precious coins in their pockets.

Under the circumstances, most banks had no choice but to suspend convertibility of their notes into specie.* The notable exception to this drastic response was in New England, where penny-pinching bankers

*The panic of 1814 was an almost perfect echo of what happened in Britain on February 1797, when the comic-opera arrival of three French frigates in the harbor of the tiny fishing village of Fishguard, Wales, set off such a panic demand for gold that the Bank of England itself had to suspend convertibility—go off the gold standard, in other words—along with the rest of the banking system. See my *The Power of Gold*, pp. 188–89.

had been characteristically zealous about limiting their note issue to what they could anticipate would be their customers' needs for specie. Elsewhere, the new arrangements did not mean banknotes were no longer convertible into gold and silver, but they were no longer convertible *on demand*: the banks could refuse to provide specie unless they had sufficient gold and silver coins on hand to cover the withdrawals, which was hardly ever. Today's world is much the same, only more so. Aside from coins for small change and a modest amount of paper currency, the virtual dollar of today consists of computer blips convertible into nothing at all except other nations' equally virtual money, much less something shiny that you can bite into, and most people never give it a thought.

Despite the shock of suspending convertibility of notes into specie, the American economy did not disintegrate and go up in smoke in 1814. Just as the economy kept perking along after President Roosevelt made ownership of gold coins illegal in 1933, and again after President Nixon's complete break with gold in 1971, people rapidly became accustomed to the new regime and proceeded to go on about their business. Like the money or not, there was no choice: money is infinitely easier and more efficient than barter for settling up transactions.

The suspension of convertibility was not about to interfere with the expansion of trade and commerce in 1814 America. The economy had been fundamentally transformed by the expansion in trade and commerce resulting from the War of 1812 and the trade embargoes that had preceded it. Production for sale was rapidly replacing production for home use. A growing volume of agricultural output was now moving to market while new factories equipped with laborsaving machinery were supplanting handicraft methods, especially in textiles.[4]* One of the

*In *Strategic Factors in Nineteenth Century American Economic History*, Claudia Goldin and Hugh Rockoff refer to the "formidable modern manufacturing sector [that] began to emerge during the first two decades of the century . . . the burgeoning manufacturing sector of the Northeast" (pp. 32–34).

most prominent innovations was Robert Fulton's steam engine and iron-smelting plant in Jersey City, which was moved over to the East River after his death, building and repairing steam vessels for Hudson River and coastal traffic as well.

* * *

Thus, in a dramatic example of the law of unintended consequences, a robust American business boom emerged from the ashes of Washington, D.C., and the panic created by the fearsome aggressions of Lord Bathurst and General Ross in August 1814. The new monetary system may have developed out of tragedy, panic, and confusion, but by liberating the banks from the restraints of specie, it converted the traditionally scowling banker into a smiling gentleman only too happy to accommodate eager borrowers by printing as many banknotes as they required.

"The plenty of money . . . was so profuse," reported a committee of the Pennsylvania legislature, "that the managers of the banks were fearful that they could not find a demand for all they could fabricate, and it was no infrequent occurrence to hear solicitations urged to individuals to become borrowers, under promises of indulgences most tempting."[5] According to Albert Gallatin, banknotes in circulation grew from $28 million in 1811 to $68 million in 1816, a compound growth rate of 19 percent a year.[6] The number of banks in the United States jumped from 88 in 1811 to 208 four years later. Pennsylvania incorporated 41 banks just in the month of March 1814. By 1818, the nation would have 392 banks.[7]

But the flood of paper money evoked nightmares among conservative people who believed the only good money was scarce money. The gush of banknotes into circulation reminded them of the collapse of the currency from overissue during the Revolution, when the expression, "not worth a Continental" was about as abusive a statement as one could make in polite company. These concerns reached Congress, where Representative John C. Calhoun warned that banks could go on granting credits ad infinitum even if they did not have a single dollar in their vaults.

Congress responded by establishing the second Bank of the United States in 1816 and opening it up for business the following year—another milestone in that remarkable year 1817—thereby reinventing an institution rudely rejected from the system six years earlier. The first Bank of the United States, Alexander Hamilton's brainchild, had been abolished in March 1811, its twentieth year of existence, in a wave of Jeffersonian antipathy to anything related to that four-letter word "bank." "I have ever been the enemy of banks," Jefferson had confided to John Adams.[8] He considered banks an even greater threat to liberty than standing armies.

There was little difference between this bank and its predecessor, except that it was larger and its charter contained about three times as many words. But by reestablishing a Bank of the United States, which would correct the conditions of the currency by both influence and example, the authorities hoped to restore order out of a system running away with itself. One of its first rules would be to refuse the notes of any bank that did not pay in gold or silver.[9]

The Bank was to be the principal depository for the federal government, which meant that is where tax payments ended up. The Bank was authorized to issue its own notes, but not in excess of its capital of $35 million, and it had to hold specie equal to at least $7 million at all times. The Bank was also permitted to accept United States government bonds from buyers of its capital stock, a step designed to improve the market and respectability of federal government debts.

The first business of this new institution was to restore specie payments throughout the banking system. Agreements to this end were reached in a complex set of arrangements, but the actual result had more appearance than substance. The failure to reach a credible restoration of convertibility was just the first in a series of steps driving the new bank in a direction directly opposite to the hopeful expectations of its original supporters, including such wealthy and conservative citizens as Stephen Girard of Philadelphia and John Jacob Astor of New York.

Instead of putting on the brakes, the Bank's first president, Captain William Jones, stepped on the accelerator. Jones was a man with strong political connections but a surprising choice nevertheless, as he had recently gone through personal bankruptcy. Prudence was not on his agenda. Right after taking over, Jones declared that he was "not at all disposed to take the late Bank of the United States as an examplar in practice; because I think its operations were circumscribed by a policy less enlarged, liberal, and useful than its powers and resources would have justified."[10]

In this case at least, Jones was a man of his word. He presided over a large increase in lending as the Bank's main office and branches in the South and west joined the rest of the banks across the country in a lending spree. By the beginning of 1818, the Bank had lent out a total of $41 million, had notes and deposits outstanding of $23 million—and a total of $2.5 million in specie in its vaults.* That was by no means all. The officers of the Baltimore branch later admitted to outright embezzlement.[11]

With the Bank of the United States setting the tune for the rest of the banking system to sing, the good times continued to spin out of control. Philadelphia State Senator Condy Raguet, an economist, wrote to the famous English economist David Ricardo, "The whole of our population is either stockholders of banks or in debt to them. It is not the *interest* of the first to press the banks [for specie] and the rest are *afraid* [to ask].

*The great London banker and philanthropist Henry Thornton, in his 1802 book, *Enquiry into the Nature and Effects of the Paper Credit of Great Britain* (reprint, New York: Augustus M. Kelley, 1991), emphasized that a government bank interested in maintaining a sound paper currency must subordinate its own profitability to restrictions in the note issue. The Bank of England was the first to adopt this principle explicitly, but the Federal Reserve and all other central banks today are managed without any direct regard for their own profitability.

. . . An independent man . . . who would have ventured to compel the banks to do justice, would have been persecuted as an enemy of society."[12]

* * *

All of this activity was great fun—while it lasted. The turning point came in the fall of 1818, when a very big chicken came home to roost. With the credit of the United States on the line, the Treasury called on the Bank of the United States to deliver $3 million in gold to the French as a payment toward the principal amount due on Jefferson's Louisiana Purchase, in accordance with the 1803 agreements. At that moment, the total amount of specie on hand at the Bank of the United States was just about $2 million. The Bank had to turn to the credit markets in London for the money, but borrowing from Peter to pay Paul solves nothing.

The bubble was about to burst. The directors of the Bank replaced Captain Jones with Langdon Cheves, a distinguished attorney with a very different view of the world from his predecessor's. Cheves was shocked at what he found. In his judgment, the Jones team had brought the Bank close to ruin without any misgivings over their irresponsible policies. He immediately decided enough was enough and cut new lending to the bone.*

Over the next two years, Cheves saw to it that the Bank's notes and deposits outstanding shriveled from $23 million to only $10 million. He accomplished this feat at the same time the Bank called upon the other banks around the country to deliver specie in return for its holdings of their notes, most of which had come in as deposits of tax revenues. By 1821, the specie reserve would be up to $8 million. But the squeeze on the economy was intense. The Bank's demand for hard cash only shifted the pressure to the commercial banks, which were now forced to press

*John Kenneth Galbraith describes Cheves as "a notably insensitive man, who may well have been what the occasion required." That was more gentle than his characterization of Jones as "a politician of questionable intelligence but proven bad judgment" (Galbraith, *Money*, p. 77).

their own customers for repayments of loans. As borrowers were driven to liquidate assets in order to meet these payments, the whole process spread pain and disruption far and wide throughout the nation.[13]

"The Bank was saved, and the people were ruined," was how one contemporary described the abrupt reversal in Bank policy.[14] Seeing his worst fears coming true, Representative John Calhoun, in a letter to John Quincy Adams, depicted the "immense revolution of fortunes in every part of the Union; enormous numbers of persons utterly ruined; multitudes in deep distress; and a general mass of disaffection to the government."[15] Twelve years later, Senator Thomas Hart Benton, still fuming over what happened, described the Bank as "the jaws of the monster. . . . One gulp, one swallow, and all is gone!"[16] After Andrew Jackson became president of the United States in 1829, he saw to it in short order that the Bank would be gone, resulting in a famous controversy with the Bank's president, Nicholas Biddle.

Contemporary estimates of distress around the country reached as high as one-third of the population.[17] In Batavia, the New York headquarters of the Holland Land Company, an observer reported on "the prospect of families naked—children freezing in the winter's storm," while others predicted that "there cannot be much ambition or hope; education will decay, and the decencies of social life be neglected."[18] Soup houses for the unemployed were established in New York as the New York Society for the Prevention of Pauperism estimated some 10 percent of the city's population was on poor relief.[19] The panic caused such devastation in the agricultural areas of Illinois, Tennessee, and Kentucky that the three legislatures created banks with the express purpose of relieving the distress of the community through low-interest loans to farmers and planters.[20] Conditions were even worse in the South, where overexpansion of cotton planting in response to the War of 1812 led to a precipitous collapse of cotton prices, from 35¢ a pound in January 1818 to 13¢ a pound eighteen months later.[21]

Tragedy stalked the propertied classes as well: the value of real and personal property in New York State declined from $315 million in 1818 to $256 million two years later.[22] Stock prices fell by 3 percent in 1818, and then by another 9 percent in 1819, a stunning reversal coming on the heels of a year when prices had climbed by 20 percent for a cumulative total of nearly 30 percent in the previous three years.[23] As a reflection of the decline in the general level of economic activity, total imports shrank from $141 million in 1818 to $75 million in 1820; the only good news from that was the end of the specie drain as imports fell off more sharply than exports.[24]

Recrimination and finger-pointing flourished, most of it using hindsight to denounce the excessive generosity of the banks in financing foolish projects and inflating people's hopes. The language was especially colorful on the subject of the "shavers and brokers" in the financial district, "who had fastened upon society like leeches, who eat out its substance and live upon its distress."[25] Tammany Hall described the nation as an "overgrown and pampered youth . . . vaulting and bounding to ruin."[26] There were those who saw all the wounds of speculation as self-inflicted and just deserts for the greedy. The cure was "to go back to the simplicity of our forefathers and exchanging . . . our dissipation for temperance and our vice for virtue."[27] A hundred and twenty years later, Russell Leffingwell of J. P. Morgan and Co. would offer the same prescription for how to get the economy out of the Great Depression of the 1930s: "The remedy is for people to stop watching the ticker, listening to the radio, drinking bootleg gin, and dancing to jazz . . . and return to the old economics and prosperity based upon saving and working."[28]

In another familiar note, complaints were widespread about the national government's failure to come to the aid of the people in distress. The spirit may have been willing, but the governmental purse was weak. In a time-honored response, John Quincy Adams pointed out that any step the government might take would only "transfer discontents [as it] propitiated one class . . . by disgusting another."[29]

The hard numbers explain the passion of the words. Manufacturing employment in Philadelphia cascaded downward from 9700 in 1815 to 2100 in 1819.[30] The price of wheat, a significant source of export earnings in New York State, plunged from $2.72 a bushel to 68¢ between 1817 and 1820, while prices of potash and hog's lard fell approximately in half.[31] Overall, prices fell 20 percent, putting enormous downward pressure on wages in the process. Although wages on the average dropped by about 25 percent, individual cases were much more severe.[32] In Massachusetts, for example, agricultural wages toppled from $1.50 a day in 1818 to only 53¢ the following year.[33]

Interest rates followed in the downward path of prices and borrowing, although with a lag. In June 1821, *Niles' Weekly Register*, a paper specializing in business and financial news, reported that an "immense capital [was] lying dead."[34] By that time, short-term interest rates were down by more than half from 1816, while long-term rates had fallen from over 6 percent to less than 5 percent.

The impact of the business depression on the cost of both constructing and financing the Erie Canal was as favorable as it was unexpected. The rapid growth in the canal's borrowing operations had a much warmer reception than anyone had anticipated, first among small savers and then among wealthy Americans and capitalists abroad. With the sea at one side, the Great Lakes on the other, the St. Lawrence River on the north, and the Hudson and Mohawk rivers running through it, New Yorkers in general and the citizens of New York City had always shown a predilection for commerce and trade. As early as 1713, a minister had declared, "The city is so conveniently Situated for Trade, and the Genius of the people so inclined to merchandise, that they generally seek no other Education for their children than writing and Arithmetick."[35] Time had served only to enhance that "Genius."

For the growing ranks of the unemployed as well as the steady stream of immigrants, the canal was one of the few places in the country where the number of jobs was growing instead of shrinking. By 1820, the canal commissioners were drawing contracts at prices 30 to 40 percent below what they had paid during the first three years of construction.[36] A writer in the *Rochester Gazette* in October 1820 described the impact of this money on the local employment situation, by pointing out that an honest laborer or farmer working on the canal would now be receiving adequate pay for his services and would be beyond the reach of "the merciless grasp of the mammoth speculators."[37]

Even the legislature was enthusiastic about the situation and authorized the Commissioners of the Canal Fund to borrow record amounts to take advantage of the abundance of labor and the low rates of interest. When the Bank for Savings lent the canal $263,000 in the second half of 1820, it had to pay $1080 for a 6 percent bond promising redemption of only $1000 in 1837, which worked out to a return of just 5.3 percent instead of 6 percent to the bank. The following year was even better, as the Canal Fund borrowed the substantial sum of $1,400,000 at 5 percent interest, and the price of these bonds in the open markets also moved above their original offering price. This was a sharp improvement from the first and second borrowings in the modest amount of $200,000 at an interest rate of 6 percent. It was an even more striking difference from the 7 percent New York State had had to pay back in 1815.[38]

These financial results were a pleasant surprise right from the start. Although the wealthy held back, people of modest means were willing to take the risk of buying the first issues of canal paper, which came out between 1817 and 1820. The $200,000 loan of 1818 drew sixty-nine subscribers, of whom fifty-one invested $2000 or less, and more than half of those invested less than $1000. Even smaller savers were able to participate in the financing of the canal through the Bank for Savings, a unique institution newly authorized, at De Witt Clinton's recommendation, by the legislature. Based in New York City and established by wealthier citizens eager to encourage thrift among "the laboring classes," the bank had been an immediate success; most of the deposits came from working men or their widows.[39] Like a beach composed of many tiny pieces of sand, the bank pooled enough of these little deposits to become one of the largest investors in canal loans. By 1821, it held almost 30 percent of the outstanding bonds.

De Witt Clinton was elated. "There is money in abundance to loan in our commercial emporium," he declared in the frothy days of 1818.[40] His observation would hold up even as the economic environment began to cloud over. In his report of 1819 to the legislature, he claimed that "in these works . . . we behold the operation of a powerful engine of finance, and of a prolific source of revenue."[41] And his secretary, Charles Glidden Haines, boasted that New York was now in a position to raise more money in a given time than the whole national government in Washington.[42]

The more risk-averse wealthy now decided they could afford to follow along the path the little investors had already cleared for them. Why not? By 1821, with nine thousand men at work on the project, the risks appeared much smaller than when small investors were taking a chance on the canal three years earlier. Every interest payment on the earlier loans taken by small investors had been met on time, even in the face of deteriorating business conditions. All the construction work on the middle section was completed in October 1820. When navigation opened up the following May, toll revenues, already trickling in since summer, would begin to accumulate. Contracts were signed or in negotiation for the entire length of the canal, while out west in Rochester the longest stone structure over water ever built in America—802 feet, with eleven arches—was under construction to carry canal boats over and across the Genesee River.

Canal paper turned into the hot new issue of the day. A number of local banks played an important role in this process as "loan contractors," which sold the Erie Canal obligations to larger investors, just as the Bank for Savings had sold them to small holders. Elkanah Watson had established one of these loan contractors in 1803, the New-York State Bank, located on State Street in Albany, a city even then filled with people seeking state office, land speculators eager to know the next area opening up, and the gamblers and other hangers-on to take advantage of the innocents. The New-York State Bank later changed its name to the State Bank of Albany, and that institution can boast today that it is the oldest bank in the United States that has stood at the same location since its inception.

Individual purchases in multiples of $10,000 became common, many bearing names that would resonate through the business history of the nineteenth century, including such astute investors as the sugar refiner Frederick Havemeyer and the tobacco merchant George Lorillard. By the end of 1822, John Jacob Astor owned $213,000 of canal loans, some of which he had bought from earlier investors, giving them a handsome profit on their investments. Langdon Cheves of the Bank of the United States, a southerner rather than a citizen of New York, who invested $45,000, was only one of many out-of-state investors beginning to participate. Soon major stock brokerage houses thought better of their earlier reluctance to offer canal loans to their customers and got busy calling on their wealthier clients to join the crowds.[43]

The result of all this was a steady rise in the amounts borrowed each year. Starting with the $200,000 loans in 1817 and in 1818, another $1.1 million dollars was raised over the next two years.[44] The canal appropriations bill of 1820 was big enough to be popularly known as the Two Million Bill. At the New York State constitutional convention of 1821, the canal's financial obligations received the state's unconditional guarantee.

The first significant purchases of canal paper by English investors appeared during the spring months of 1822. And again, why not? Liquid capital was piling up, as the defeat of Napoleon in 1815 led to the end of the British government's massive borrowings. In any case, canals had been favorite investments for English investors for more than twenty years, and the stocks of many canal companies had turned out to be spectacular performers, paying huge dividends as well as soaring appreciation in their prices.[45]

The Erie Canal loans became as irresistible to the British as high-tech stocks were to investors in the late 1990s, only they were a lot less risky. The English press was full of tempting pieces of news about progress on the Erie Canal, with reports that "the junction of the American Lakes with the Atlantic Ocean, calculated to improve prodigiously the commerce of New York, goes on nobly to its completion." Or that New York City was about to become "the London of the New World."[46] With 220 miles of the canal navigable and earning tolls in 1822, there were no more lingering doubts about the canal's huge scope, the engineering challenges, the complex administrative tasks, and the ability to finance and hold costs to original estimates.

The Canal Fund borrowed a total of just under $2 million from 1822 to 1824, which completed its financing needs. The available supply of canal bonds was now too small to satisfy rising British demands. By 1823, the British were gobbling up the new issues and just as voraciously paying a premium to buy positions in earlier issues held by those Americans who had accumulated canal paper earlier in the game. By 1829, foreigners would own more than half of the canal's outstanding obligations of $7.9 million.[47]

Although these borrowings were enormous sums for their times, the burden on New York State was smaller than the bare numbers would suggest. While the Canal Fund was issuing obligations totaling $7,896,150, New York State's general fund was operating at a surplus. The state managed to pay down its direct debts, from $2.7 million outstanding when construction on the canal began in 1817, to zero in 1825 when construction was complete. No wonder the full faith and credit of New York State made the canal bonds such a choice investment. The entire process would turn out to have created the foundation for the sophisticated capital market of New York, preparing Wall Street for future financings no one could begin to visualize in the early 1820s—the railroads, the Civil War, and the huge industrial debt issues that followed.[48]

* * *

With the construction and financing of the canal moving along so smoothly, its future should no longer have been a matter of doubt. Yet old controversies were still festering, and the canal continued to be the focal point of the political wars raging with sustained intensity throughout this whole period.

In his message to the legislature in 1819, De Witt Clinton predicted what lay ahead when he warned that "although the efforts of direct hostility to the system . . . will in future be feeble . . . there is great reason to apprehend the exertions of insidious comity [and] I consider it my solemn duty to warn you against them. . . . Good men will receive the countenance of persons of a different description, who in furtherance of selfish designs, will strive to destroy the great fabric of internal improvements."[49]

Clinton may have exaggerated when he cautioned against "persons [who] will strive to destroy the great fabric of internal improvements," but he was correct in his intuition that the future of the canal, and of his own political destiny, were far from settled.[50] There were to be no prisoners taken in the looming battles for control.

CHAPTER 13

RUDE INVECTIVE

It is difficult to separate De Witt Clinton's political achievements from the Erie Canal, because he made the canal the centerpiece of his policies and goals. But then the canal inevitably became the lightning rod for the political hostility he faced for so many years. Clinton's political enemies never lost their zeal for taking aim at Mr. Erie Canal. Had the opposition had its way at critical moments, the canal would have ended up shorter and with a different structure from the engineering masterpiece that was finally realized.

At the beginning, Clinton's enemies were determined to prevent the canal from ever becoming a reality, but Martin Van Buren's belated decision in 1817 to support the canal was the end of that effort. The opposition had to change their tactics even if their strategy remained to eliminate De Witt Clinton from the political scene. By the end of 1819, when Elkanah Watson was declaring, "It was impossible for stupidity itself [not] to stretch their opaque minds from Erie to the ocean," those "opaque minds" were managing a remarkable transformation in the character of the debate. Watson was right. If it was stupid to continue the fight against the canal, the opposition would claim *they* were the ones with the true faith and greatest concern for the enterprise. They had insisted all along that Clinton had no genuine interest in the canal except as an instrument to promote his own political future.

Clinton fought back by maintaining—as *he* had insisted all along—that only he was entitled to wrap himself in the glorious garb of the Erie Canal. His vision and leadership, after all, had motivated and shaped the whole strategy. He was the one who had taken the career risks to see it through from the very beginning.

All too often, Clinton failed to outmaneuver his opponents, despite the superiority of his arguments. His vulnerability was not in his defensive strategy or in the unassailable records of his achievement. Events would prove that his liabilities were in his personality—his disdain toward enemies and often to friends as well, his unwillingness to compromise, and his intolerant manner to those who disagreed with him.

In a remarkable manner, Clinton's diary of the expedition west in 1810 reveals his lack of a common touch. He seems to be more interested in the life of the birds and the fish he sees than in the personal lives of the farmers he interviews about the cost of their land and the prices they receive for their crops. He promoted free schools and the manumission of slaves in New York, but the diary shows a terrible and uncontrollable contempt for the lower classes he encountered in small-town inns along the future route of his beloved Erie Canal.

Power for high purposes consumed him; politics he disdained. Most politicians simply did not like him; many could not abide him despite his wide popularity among the public. Jabez Hammond, a longtime member of the legislature and able historian of the times who was generally sympathetic to Clinton, could not avoid complaining that Clinton always pursued noble and magnificent ends but fell short when it came to providing the means to accomplish those ends.[1] An acquaintance put it more sharply, and directly to Clinton: "And let me tell you, Sir, if no one else has the candour or boldness to say it (I mean among your friends), that the charge of a cold repulsive manner is not the most trifling charge that your political enemies have brought against you—you have not the jovial, social, Democratical-Republican-how-do-you-do Suavity . . ."*

Clinton was deeply concerned about his country's passion for political factionalism, even though, in spite of himself, he contributed his fair share of it. In a world where most towns were little and entertainment sparse, the great game of political brawling was a reliable source of entertainment. Some banks in New York and Albany regularly denied credit to Jefferson's supporters.[2] The use of language was especially noxious. During the election campaign of 1800, for example, Federalist newspapers predicted that the election of Jefferson would cause the "teaching of murder, robbery, rape, adultery and incest."[3] In a long and rambling lecture before the Literary and Philosophical Society of New York in 1814, Clinton bewailed these descents into the gutter to attack political opponents: "Our ingenuity has been employed, not in cultivating a vernacular literature, or increasing the stock of human knowledge; but in raising up and pulling down the parties which agitate the community. . . . The style of our political writings has assumed a character of rude invective, and unrestrained licentiousness, unparalleled in any other part of the world, and which has greatly tended to injure our national character."[4]

Clinton was by no means immune from the criticisms he directed at others. In true Shakespearean style, he was both admirable and despicable, more brilliant, more foresighted, and more effective than his enemies and yet he himself wreaked the greatest damage to his hopes and dreams. His airs of cultural and intellectual superiority were often at odds with the gusto with which he splattered insults at his assailants, as we have seen when he had described his former friend and ally Governor Tompkins in February 1816 as "destitute of literature, science, and

*John Brennan's letter, postmarked September 23, 1823, is quoted by Dixon Ryan Fox in *The Decline of Aristocracy in the Politics of New York* (p. 200). Fox, who is clearly no fan of Clinton, goes on with an extended recital of nasty commentary about him for another ten pages, to say nothing of interspersed observations and quotes of this character throughout his book.

magnanimity—a mere creature of accident and chance, without an iota of real greatness." And there would be more to come.

But Clinton did not use language lightly. His rich and colorful fulminations against his opponents bring to mind John Maynard Keynes's advice that "words ought to be a little wild for they are the assault of thoughts on the unthinking."[5]

* * *

From the start, Clinton's most relentless enemies were concentrated in Tammany Hall, where animosity toward him had roots reaching back at least to 1802. The most remarkable feature of the battle was not in its intensity but in its irrepressibility. There was almost no letup until Clinton put *finis* to the matter in 1828 by dying while he was still governor. Although the conflicts between Clinton and Tammany usually appeared to be over matters of policy and substance, the heart of the matter was a bitter contest over power and patronage, seasoned with the envy and frustration Clinton provoked throughout his career.

The Society of St. Tammany had been founded in 1786 by an ex-soldier named William Mooney, keeper of an upholstery shop on Nassau Street. The word "Tammany" was an Americanization of a larger-than-life and liberty-loving seventeenth-century Indian chief known as Tamamend, whose wigwam is reputed to have stood on the grounds of what would one day be Princeton University.

Among the elaborate goals set forth by Mooney, the key phrase was Jeffersonian: "to sustain the State institutions and resist a consolidation of power in the general Government."[6] If Jefferson was Tammany's hero, the enemy was Alexander Hamilton and all he stood for: a dominant central government (Hamilton had favored having both the president and the Senate elected for life), support of the clauses of the New York State constitution limiting the right to vote for governor or state senators to male owners of debt-free freeholds of £100 or more, and the removal of all political and legal disabilities of former Tories, or supporters of Great Britain.

As populists and anti-elitists, the members of Tammany went to great lengths to distinguish themselves from the patrician societies like the Society of Cincinnatus for ex-officers of the Revolutionary Army. Tammanyites sported aboriginal designations and customs by dividing themselves into Indian tribes and adopting titles ranging from Grand Sachem for their officers and just plain Sachem for the board of directors down to Wilkinsie for the doorkeeper. They even used the Indian calendar, with patriotic emendations: July 1800, for example, became "Season of Fruits, seventh Moon, Year of discovery three hundred and eighth; of Independence twenty-fourth."[7] Their first home, Barden's Tavern, was dubbed the Wigwam, a name that clung to Tammany's quarters forever after. But such enthusiastic evocation of the Indians was still not enough to separate the Tammanyites from Hamilton and his ilk. The frontier spirit was equally important, as exemplified by their fondness for hats with flowing bucktails, or the tails of male deer. So devoted were they to these hats that the Tammany-driven forces opposing Clinton within his own party came to be known as the Bucktails.

With the passage of time, Tammany became increasingly Irish and pro-Catholic, an attitude that would have been unthinkable in the early 1800s. But the spirit and the view of themselves as partisans for the common man persisted throughout their entire history, regardless of religious considerations.

Early activities at Tammany were mostly patriotic celebrations in the grand style, with prolonged speeches about the glory of the American Revolution and, on occasion, about the revolution of the French as well. The celebrations ended with sumptuous dinners and a drawn-out sequence of toasts in honor of Citizen Jefferson and their "fellow-fallible mortals," as well as toasts in opposition to "Ambition, Tyranny, Sophistry, and Deception."[8] The greater the number of toasts, the greater the excuse to keep on drinking. After about a dozen years at Barden's Tavern, Tammany relocated the Wigwam next door to Martling's Tavern at the corner of Nassau and Spruce streets, then at the very outskirts of the city. The building was so rundown many people referred to it as the Pig Pen.*

By 1800, the Tammanyites were in the process of changing from a purely social society into a private political club to support Jefferson's Republican Party and—a more controversial goal—to whip up enthusiasm for the French Revolution. Although through all the years Tammany never wavered from its opposition to the conservative political parties, its real aim was to influence elections so that officeholders would be indebted to them, would do their bidding, and would, in one way or another, arrange for payoffs into the pockets of the Tammany leaders. None of these activities was ever allowed to interfere with the ceremonial fun and games the members of Tammany enjoyed so much.

* * *

This shift in focus was accelerated when former New York State senator and assemblyman Aaron Burr joined in 1800, attracted by both the anti-Hamiltonian flavor of the society but also by his vision of Tammany as a potentially powerful political base. Burr, then forty-four years old, had served on Washington's staff in the Revolution—until he so antagonized Washington that he was transferred to other duties. He developed a prospering law practice in early postwar years but entered public life in 1789 when Governor George Clinton appointed him attorney general for New York. Two years later, he won a seat in the U.S. Senate by defeating General Philip Schuyler of Western Inland Lock Navigation Company fame. Schuyler, it might be recalled, was Alexander Hamilton's father-in-law. Hamilton and Burr were enemies from that point forward. But when Burr failed to win reelection to the Senate in 1797, he turned to the political scene in New York—and Tammany Hall—at just about the same time as his participation in the presidential election of 1800.

*Tammany took care of Sachem "Brom" Martling in his old age by securing his appointment as Keeper of City Hall, a job with no responsibilities except to show up whenever the spirit moved him.

Henry Adams, one of Burr's few firm admirers among historians, describes him as "an aristocrat imbued in the morality of Lord Chesterfield and Napoleon Bonaparte" and excused his frequent lack of rectitude by observing, "Great souls care little for small morals."[9] Another author tells us, "It was once observed of Aaron Burr that his sole claim to virtue lay in the fact that he himself had never claimed it."[10] It may be an overstatement to say that Burr invented the American political machine, but there is no doubt he understood how to convert Tammany's boisterous dinners, parades, and drinking bouts into a potent political instrument to advance his own objectives.[11]

Burr's style was to prefer the action behind the scenes to performing out front. He became Tammany's unquestioned leader, but he never took the title of Grand Sachem and seldom showed up at Martling's Tavern. He delivered no long rhetorical speeches and left publishing great essays to others. While in office he never originated any great measure of political importance. His correspondence is colorless and lacking in any significant insights about the world around him. Yet Burr had white-hot political aspirations and no inhibitions about achieving them.

While the ceremonies and good times carried on by the rank-and-file members of Tammany continued to give an impression of simplicity and celebration of the common man, power was the Burr contingent's single objective. Banks can finance power, and Burr's success in launching the Bank of the Manhattan Company in 1799 gave him just the structure he needed for lending money to favorites and financing whatever electioneering he deemed appropriate. In so doing, he was only following in the footsteps of Alexander Hamilton himself, who employed the banks over which he had influence—the Bank of New York and the Bank of the United States—in the same way. Hamilton had no inhibitions about employing his banks to further his political ambitions and to deprive his political enemies of financial accommodations.

But Burr's Bank of the Manhattan Company claimed it was ready to lend to any prospective borrower worth the risk involved. Burr also brought most of his coterie along with him, most notably his friend Matthew Davis, the prototype of the local political boss and later Burr's first biographer. Davis pioneered Tammany's methods of holding dummy meetings in the city's wards, influencing the nominations of favored candidates, and passing high-sounding resolutions designed to influence votes in the wards and even throughout the state. In time, Tammany would be busy rounding up its supporters at election time to vote for the Tammany candidates—even when some of the "supporters" may not have been technically qualified to vote.

As these activities became increasingly effective in influencing the election results, they also cleared the path for the patterns of corruption and unabashed stealing from public funds for which Tammany became famous over the years. Few people appear to have been fooled by the outward appearances of Tammany activities, even at the start. Gustavus Myers, author of an exhaustive history of Tammany first published in 1901, reports that a long line of Tammany leaders was involved in a wide variety of public and private frauds beginning in 1799 and 1805–1806 and extending to the later decades.[12] In fact, the index of Myers's history shows far more citations under "corruption" than for any other single subject or proper name mentioned in his book. The 1807–1808 edition of *Longworth's American Almanac* describes Tammany as having "a constitution in two parts—public and private—the public relates to all external or public matters; and the private, to the arcana and all transactions which do not meet the public eye."[13] No matter how frequently the Tammanyites were caught with their hands in the till over the years, they seem to have kept right at it with irrepressible enthusiasm.

Burr hoped to be president of the United States one day, and he joined Tammany just as he was getting set to run in the presidential campaign later in 1800. When the votes in the Electoral College were counted, Burr and Jefferson were tied for first place, with the incumbent President John Adams not far behind, and Thomas Pinckney and John Jay trailing behind. Although most people (probably including Jefferson himself) expected Burr to step down and accept the vice presidency out of respect to the author of the Declaration of Independence, Burr refused. The tie then had to be resolved in the House of Representatives. Hamilton used all his influence to swing the vote in Jefferson's direction, even though he and Jefferson were barely on speaking terms and had been on the opposite side of almost every political issue. As he wrote to Gouverneur Morris, "[Burr's] elevation can only promote the purposes of the desperate and the profligate."[14]* On the thirty-fifth ballot the votes finally went for Jefferson and left Burr stuck in the subordinate role. It was Hamilton's outspoken support for Jefferson as the lesser of two evils in this contest—to say nothing of the bitter competition between Hamilton and Burr in the banking business in New York—that led ultimately to the duel between them in 1804 that left Hamilton nearly dead on a New Jersey field. His seconds rowed him back across the Hudson to New York, but he died about thirty-six hours after receiving the fatal wound.

Tammany was loyal to Burr to the end. Two leaders of the society accompanied him to the duel. Another, John Swartout, was at Burr's home to welcome him upon his return from the duel; Swartout had been among Burr's chief aides when he was vice president. But Burr had no choice but to leave New York at that point. Despite his absence, his followers would control Tammany Hall for another thirty-five years.

In 1802, Burr and the Clintons collided. It was two years after the election but two years before the duel when the *American Citizen*, a New York newspaper well known for serving as a voice for De Witt Clinton, accused Burr of being a traitor to the Republican Party by having refused to yield to Jefferson without a vote in the House. From that point

*On the other hand, John Marshall, then Adams's secretary of state, had this to say about Jefferson, with a curiously meticulous use of grammar: "His foreign prejudices seem to me totally to unfit him for the chief magistry of [the] nation. . . . He will increase his personal power [and] sap the fundamental principles of the government" (Roger Kennedy, *Burr, Hamilton, and Jefferson*, p. 54, citing Alfred Beveridge, *The Life of John Marshall* [Boston: Houghton Mifflin, 1916], vol. 2, p. 537).

onward, De Witt and his uncle George Clinton, then governor of New York State, were determined to dislodge Tammany and all its influence from the Republican Party. George Clinton as governor got a running start on this objective when Jefferson worked through him rather than through Burr in deciding who would receive federal appointments from the Republican Party in New York. Yet the Clintons would never complete the job of erasing Tammany from the political arena.

In the short run, the Clintons were able to take over control of the Bank of the Manhattan Company by influencing just enough share owner votes to boot Burr and John Swartout off the board of directors. Not long after, De Witt Clinton—despite his expressed aversion to "rude invective"—referred to Swartout in the course of conversation as "a liar, a scoundrel, and a villain" because Swartout had accused of him of bad character.[15] When Swartout demanded satisfaction, Clinton offered to settle with mutual retractions on each side, but Swartout was not to be talked out of the duel.

Swartout's determination in this matter was more routine for his own time than it sounds to people living in the twenty-first century. At a time when political allegiances were just beginning to form and party politics was in its infancy, personal reputation, the gossip channels, and the uninhibited language in the press often played dominant roles in the formation of alliances and enmities. Duels were an infrequent but acceptable climax to these struggles for power and attention. Many protagonists responded in the spirit of Lord Chesterfield's view of the matter: "There are but two alternatives for a gentleman; extreme politeness or the sword."[16]

As New York prohibited dueling, Swartout and Clinton met in New Jersey on a late July afternoon of 1802, having rowed over in the company of their seconds and a surgeon. Although Clinton declared his preference for standing face-to-face, Swartout insisted on back-to-back, turning, and then firing. Both missed on the first exchange. Clinton sent word he was satisfied and hoped they could now renew their old friendship. Swartout refused. The whole procedure was repeated twice, with Clinton's coat receiving one of the bullets. On the fourth shot, Clinton's bullet hit Swartout in the leg. After the surgeon had removed the bullet, Swartout rose and demanded that the duel must continue. Clinton announced he bore Swartout "no enmity," but Swartout remained adamant. Even when the sixth shot landed in his leg, he was still demanding another chance to kill his opponent. Clinton refused to fire again.[17]*

Years of unbounded acrimony and struggle now lay ahead for De Witt Clinton, punctuated by a few brief efforts at reconciliation. They all ended in failure. There were too many prizes of patronage, influence, and prestige, and too many memories of harsh recriminations, for either side to back away or compromise. Henry Adams, an implacable critic of Clinton, asks why "no one ever explained why Burr did not drag De Witt Clinton from his ambush to shoot him, as two years later he shot Alexander Hamilton with less provocation."[18] Aaron Burr may have been the match that lit the fuse, but the struggle would continue on merrily long after he was just a memory.

* * *

In 1818, the peak of the business boom and the first full year of construction on the Erie Canal, some three thousand men were at work on the canal and the commissioners were projecting completion of the middle section by the end of the work season the following year. But none of this impressive development inhibited the New York City delegation to the state Assembly from denouncing Governor De Witt Clinton and his enthusiasm for internal improvements. They even continued to predict failure for the whole project, and the financial ruin of the entire state as a consequence.

Observing the temper of the legislature, Jabez Hammond warned

*John Swartout survived his wounds in fine style. He lived for another twenty-one years.

Clinton he was in deep trouble, even among members of his own party. Clinton laughed it off. When a few of Clinton's more sober opponents sought a compromise in some of the disputes raging among them, and Senator Hammond again appealed to Clinton to be more responsive to what was going on, he found Clinton "indifferent, and disinclined to aid in carrying into effect the project. . . . From that moment my confidence in Gov. Clinton . . . was gone. I have never regained it."[19]

Clinton had also developed the habit of building links to the Federalists. Even though he was reluctant to appoint Federalists to state positions, his shrinking support from his own party forced him to reach out to the other party for fear of losing all control over the legislature. The more he reached out, the more intense the attacks from his fellow Republicans. Samuel Hopkins Adams's line in his 1944 novel—"Clinton, the federal son-of-a-bitch"—caught the spirit of the times.

The situation left Clinton in the unhappy position of facing a well-organized majority against him in the state senate. Van Buren was inspired to depict Clinton "on a giddy eminence on which the difficulties of preservation are so great & distressing that many men would prefer political prostration as a choice of evils."[20] Although Clinton would respond that Van Buren was "an arch-scoundrel . . . a confirmed knave," the Federalists whom Clinton had failed to recruit to his cause were as contemptuous of him as the Van Burenites.[21] "He is the political Ishmael of our times," wrote one Federalist in the *New York Post*, "a sort of *political pirate* sailing under his own black flag and not entitled to use that of either party."

The tenor of the controversy was ridiculous, whirling around personalities rather than issues. There was no doubt the canal enjoyed a strong majority in the legislature even if Clinton did not. Support for the canal was so apparent that even the Bucktails could see their negative position was getting them nowhere. "In a twinkling of time," as one of Clinton's allies described it, they abruptly reversed themselves and jumped into the pro-canal camp.[22] After that, points of difference between the Bucktails and the Clintonians over matters of substance were minimal.

In the legislative session of 1819, the Bucktails took advantage of their new championship of the canal to find fresh reasons for making trouble for Clinton. They began with what appeared to be just an extension of their usual opposition to the canal by attempting to block any work on the eastern or western sections before the middle section was complete. Their primary goal, they bragged, was to sustain popular support for the canal, and they justified this negative position by claiming the voters might desert the canal without a full demonstration of its feasibility, the skills of the construction crews and their managers, and the adequacy of the toll revenues. The last was a particularly clever barb: no tolls would be collected until the following year. Clinton managed to muster enough support to beat back this proposal, but the opposition was both vocal and stubborn.

After this, Clinton finally recognized serious trouble was brewing. He urged the commissioners to lose no time in signing all the necessary contracts for the sixty-four miles from the western end of the middle section out to Rochester and for the twenty-four miles of the eastern section between Utica and Little Falls. "By operating in both directions," Clinton contended, "a solemn pledge is given of our determination to finish the whole canal."[23] "Whole" was the operative word.

But letting contracts was no assurance of legislative appropriation of funds when the time came. The stretch from the Seneca River to Rochester appeared to be secured, but once the work to Rochester was complete, a new struggle would open up over the final western leg, between Rochester and Lake Erie at Buffalo. Indeed, the southern counties in particular could see no benefit to them from all this expenditure of the state's money. In addition, Peter Porter and his friends were still agitating for the route across Lake Ontario westward to the mouth of the Niagara River, where Porter's extensive landholdings would enjoy greatly enhanced values. It would never be easy going.

* * *

The next episode in this tangled story turned into a major setback for Clinton. Joseph Ellicott, resident agent in the west for the Holland Land Company, had resigned from the canal commission during the summer of 1818 on the basis of poor health. Governor Clinton nominated Ephraim Hart, one of his strongest supporters, to replace Ellicott. It was an unfortunate choice. Hart was not only anathema to the Federalists whose support Clinton so badly needed, but he was also generally unpopular for his rough manners, indiscretions, and intolerant attitudes. The Bucktails seized the opportunity to propose Henry Seymour, one of Clinton's most bitter opponents but evidently a man who was courteous in his relations with opponents as well as friends and therefore popular in both political parties. Early in 1819, the Senate chose Seymour, by a one-vote margin.

Despite the narrowness of the vote, the Bucktails now held more than half the votes on the canal commission, depriving Clinton of the control he had exercised—and cherished—from the very beginning. The immediate consequences of Seymour's victory were expressed in the Bucktail newspaper, the *Albany Argus*, on the morning following the vote: "A majority of the canal commissioners are now politically opposed to the governor, and it will not be necessary for a person who wishes to obtain employment on the canal as agent, contractor, or otherwise, to avow himself a Clintonian."[24] The required avowals did not need spelling out. Rage overcame Clinton once again. Among other fulminations, he addressed Samuel Young, one of the Bucktail commissioners, as "a blackguard [guilty of] audacious calumnies."[25] Young had been Clinton's scholarly colleague on the commission since 1816 and had delivered the speech at the ceremony in Rome in 1817 when the first shovelful of earth marked the beginning of the canal's construction. But Young was also about as intolerant as Clinton of those who disagreed with him, and he constantly cast doubts on their integrity and intellectual capabilities.[26]

There was one ironic note in Clinton's favor at this moment. General Andrew Jackson arrived in New York City for a visit, and Tammany Hall threw a dinner in his honor. After twenty-two toasts to a long list of American heroes, the dinner chairman finally asked Jackson to make his contribution. Raising his glass, Jackson proposed "De Witt Clinton, the enlightened statesman, governor of the great and patriotic state of New York." Was he aware of the bomb he was about to drop? The next day's *New York Post* described the resulting pandemonium in a ditty:

> The songs were good, for Mean and Hawkins sung 'em.
> The wine went round, 'twas laughter all and joke,
> When crack! The General sprang a mine among 'em,
> And beat a safe retreat amid the smoke.[27]

* * *

The elections for the legislature later in 1819 left Clinton even further behind in the conflict. It was an especially ugly campaign for the governor, coming at a time when he had just married for the second time and was trying to settle down to a happier life.[28] Although the Bucktails failed to overcome Clinton's majority in the Assembly, their gains cut deeply into his margin, while the Senate was already under Bucktail control. And now the gubernatorial election of 1820 was approaching.

The Bucktails chose Daniel Tompkins as their candidate, after he acceded to Van Buren's insistence he was the only man who could inflict a climactic defeat on the great enemy, De Witt Clinton. It must have taken some strong arm-twisting, because Tompkins would have to retire as vice president of the United States under James Monroe. There was no doubt Tompkins was immensely popular in the state, but his attachment to the Erie Canal was doubtful. As governor, he had displayed no enthusiasm for the canal. While vice president, he had attended the meeting of the Council of Revision in 1817 and had pressed the members—without success—to vote against the canal because of his certainty of another war with England within two years. In addition, he was still trying to respond to accusations about the mysterious disappearance of state funds while he was governor during the War of 1812. Tompkins's main attraction, aside from his long experience in government, was the contrast between his steadfast geniality and Clinton's quick temper and excessive sensitivity.

The election campaign turned largely on which party had displayed the most consistent devotion and loyalty to the Erie Canal. Nobody spared words. The *Niagara Patriot*, a Bucktail newspaper, accused Clinton of being "sycophantic, disgusting, and dishonest . . . illiberal [and] undignified" in his efforts to deny the Bucktails the credit they were due.[29] The Clintonian press made capital of the refusal of Bucktail commissioners Young and Seymour to sign the commissioners' report for 1819, accusing them of unremitting malice toward the canal and secretly aiming to destroy this magnificent project. Tompkins worked the southern tier of the state, where opposition to the canal was strongest. Clinton reminded the growing number of Irish of his concern for their welfare and Tammany's outspoken antipathy toward immigrants.

When election day finally arrived at the end of April, Clinton emerged as victor with a winning margin of 1457 votes out of a total of some 90,000. Most of this paper-thin majority came from the counties along the line of the canal. Nonetheless, the Bucktails had the satisfaction of achieving control of both houses of the legislature.[30]

Van Buren was jubilant. In a letter a month later, he wrote, "We have scotched the snake, not killed it. One more campaign & all will be well."[31] His first sentence was beyond dispute. His second sentence, like most forecasts, would turn out to be little more than a simple extrapolation of the recent past. Indeed, Clinton, making quite the opposite forecast, declared to a friend, "The Bucktails have risen to fall like Lucifer, never to rise again."[32]

The Bucktails opened the legislative session in November 1820 by passing, with no votes opposed, a canal appropriation of $1,000,000 for each of the next two years to supplement the $600,000 annual appropriations already in place. This was the Two Million Bill enacted at the worst of the business depression, when the commissioners and the legislature were under pressure to move ahead as rapidly as possible because of the sharp reduction in the cost of labor and in interest rates. The legislation also contained these critically important words: "This state can never enjoy a tenth part of the advantages of the Erie canal, till the tide of inland commerce, of which it is to be the channel, is permitted to flow, without a mile of portage, from the great lakes to the Atlantic."[33] The state legislature was now committed to completing the canal as originally planned, instead of stopping short somewhere around Rochester.

Money matters, and the Two Million Bill put the Bucktails clearly on the side of the canal. But there was a second section of the bill, which provided for increasing the membership of the canal commission by one, and that would give the Bucktails an opportunity to increase their majority on the board. The legislation also stipulated that, at "the pleasure of the two houses," any commissioner could be removed from office by a concurrent resolution. This provision gave the Bucktails a stranglehold on the management of the canal—and all the patronage that came with it—for as long as they could maintain their majorities in the legislature.

Repelled by this threat to his power, Clinton's instinct was to veto the bill. That is precisely what the Bucktails were hoping for. A Clintonian state senator named Gideon Granger somehow managed to persuade this stubborn man to change his mind. The oversized appropriation of money was the most important part of the legislation, Granger argued, because it essentially assured the completion of the canal in the form Clinton had urged from the very beginning. His entire reputation and political future rested on that achievement. The price of living with Bucktail commissioners was worth it.

Clinton agreed. He signed the bill. The opposition was stunned. While they had provided tangible proof of their support of the canal, they still found themselves guaranteeing Clinton's place in history whether they liked it or not. Since the passage of the first canal bills in 1817, no turning point in the entire history of the canal would match this one for high drama. Clinton now felt free to make his own prediction: he told the legislature the canal would be finished by 1823 "at the farthest." Myron Holley, an early enthusiast for the canal, declared that "the dearest hopes of its friends . . . precarious till this Spring . . . may be regarded as sure and steadfast."[34] The Bucktails tried to join in the celebration, avowing they had finally proved beyond dispute the villainous deceit of Clinton's repeated accusations that they were the enemy of the canal. Clearly, they claimed, they were its most constant and energetic supporters.

The Bucktails had more on their agenda than the canal. They were eager to convene a constitutional convention that would, among other things, reduce the governor's term from four years to two and move election day to November. As the arithmetic worked out, Clinton's new term would end in January 1823, fifteen months before the date provided for under a four-year term. Sensing what lay ahead, and eager to gain political support from the thousands of new settlers in the west, Clinton fought to delay the convention until the results of the census of 1820 were available. But the Bucktail majorities in the legislature pushed ahead over his objections.

The convention was for the most part more high-minded than this. It put the full faith and credit of the state of New York behind the Canal Fund's obligations into perpetuity. The Council of Appointment and the Council of Revision were abolished, removing major legislative roadblocks to the power of the governor. Finally, the new constitution included two important provisions for the advancement of democracy in the state. First, it extended the suffrage by eliminating just about all property qualifications for both voting and holding office—except for African Americans, who still had to own and pay taxes on at least $250 worth of property, a stiff standard for them to overcome (they would be totally disenfranchised in the convention of 1846). Second, following Clintonian proposal, the new constitution provided for direct popular election of the state's electors in presidential campaigns instead of by the state legislature.

* * *

Clinton was clearly on the defensive all the way from Ellicott's resignation from the canal commission in 1818 right through to the end of his first term as governor in January 1823. On the occasion of his final annual address of that term to the legislature on January 2, 1822, Clinton pleaded for unity and an end to vituperation: "Whatever diversity of opinion may exist, I am persuaded that we will all cooperate . . . in cherishing a spirit of conciliation and forbearance, and in cultivating that respect which we owe to each other and to ourselves." The legislature's response, through a committee resolution, was blunt: "The custom of delivering a speech by the Executive to the Legislature at the opening of the session, and of returning an answer to the same, is a remnant of royalty, not recommended by any considerations of public utility, and ought to be abolished."[35]

A few redeeming moments remained, such as his diary note in May 1822 during a trip on the canal that he "saw a great number of ascending and descending vessels, among the rest 'Lady Clinton.'"[36] But there was a much more memorable event ahead. Late in December 1822, he received a letter from Thomas Jefferson:

> I thank you, Dear Sir, for the little volume sent me on the natural history and resources of N. York. it [*sic*] [is] an instructive . . . and agreeably written account of the riches of a country which our great canal gives value and issue, and of the wealth which it creates from what without it would have had no value. Altho' I do not recollect the conversation with Judge Forman referred to on page 131, I have no doubt it is correct, for that I know was my early opinion, and many, I dare say, still think with

> me that N. York has anticipated by a full century the ordinary progress of improvement. This great work suggests a question, both curious and difficult, as to the comparative capability of nation to execute great enterprises. It is not from a greater surplus of produce after supplying their own wants, for in this N.Y. is not beyond some other states; is it from other sources of industry additional to her produce? This may be,—or is it a moral superiority? a sounder calculating mind, as to the most profitable employment of surplus, by improvement of capital instead of useless consumption? I should lean to the latter hypothesis were I disposed to puzzle myself with such investigations, but at the age of 80 it would be an idle labor, which I leave to the generation which is to see and feel its effects.[37]

By the time Clinton's term ended in January 1823, he was exhausted from the struggle. His financial affairs were also in disarray, to a point where he was borrowing from friends and even forced to sell shares in his library, valued at $20,000.[38] He declined to run for reelection and returned to home life and his many activities in the sciences and the arts.

But he could not bring himself to go all the way to abandon politics for the finer things of life. His attention was glued to the gubernatorial campaign between a printer named Solomon Southwick and the Bucktail choice, Justice Joseph Yates, who had been on the Council of Revision when Vice President Tompkins had nearly succeeded in blocking the Erie Canal in 1817. Clinton repeatedly predicted Yates's certain defeat until the very last moment. Yates won in a landslide, getting 98 percent of the vote.[39] Even then, Clinton asserted to a friend that "if I had been a candidate, I would have been reelected governor."[40]

Clinton did decide to remain as a commissioner of the Erie Canal, but further cruel bumps lay ahead even there. The Clinton story and the canal story were moving in opposite directions. At the canal, all the news was wonderful. The first tolls had been collected in July 1820. By the end of the year, nine miles of the western section had been filled with water

Ten-term mayor of New York City and three-time governor of New York, De Witt Clinton provided the strongest leadership for building the Erie Canal. While under construction, the canal was known as Clinton's Ditch, and for good reason.
(*De Witt Clinton* by Samuel Morse. The Metropolitan Museum of Art. All rights reserved.)

At the Wedding of the Waters, the elaborate ceremonies in New York harbor to celebrate the completion of the Erie Canal in 1825, Governor De Witt Clinton poured a cask of water from Lake Erie into the waters of the Atlantic Ocean.
(*Marriage of the Waters* mural by C. Y. Turner. De Witt Clinton High School in New York City.)

George Washington was the first to foresee the necessity of a waterway providing a commercial link between the population of the states along the Atlantic Ocean and the people moving west beyond the Appalachian mountain chain.
(*George Washington* by Gilbert Stuart. Copyright © The Frick Collection, New York.)

Thomas Jefferson refused federal financing for New York's Erie Canal, despite promises made earlier. In 1809, he declared that the proposal to build an artificial waterway across the state was "little short of madness to think of . . . at this day."
(*Thomas Jefferson* by Rembrandt Peale. The White House Historical Association; White House Collection, 55.)

Shortly before 1550, Leonardo da Vinci designed the first lock gates in a V pointing upstream when closed, so that the pressure of the water flow would keep the gates tightly shut.
(*Canal Lock* by Leonardo da Vinci. V&A Images, Victoria and Albert Museum.)

Gouverneur Morris, an author of the Constitution of the United States, predicted in 1800 that "hundreds of ships will in no distant period bound the billows of those inland seas" in the interior of New York State. (*Gouverneur Morris* by Ezra James. New-York Historical Society, accession #1817.1.)

In 1788, Elkanah Watson was among the first to explore western New York State with a view to establishing canal communication between the Great Lakes and the Hudson River. (*Appleton's Cyclopedia of American Biography*, 1887. Image courtesy of Stanley K. Kos-stanklos.com.)

President James Madison promised to promote federal financing for public works but did not deliver on his promises, even after the War of 1812 revealed the urgent need for improved transportation between the Atlantic Ocean and the Great Lakes. (Library of Congress.)

While Secretary of the Treasury under Jefferson, Albert Gallatin drafted a report concluding that federal financing of great public works—including a canal across New York State—was essential to promote economic growth. (*Albert Gallatin* by Rembrandt Peale, from life, 1805. Independence National Historical Park.)

Peter Porter was a politician and an aggressive entrepreneur in the area of Niagara Falls who supported a waterway to the west—as long as it would fatten his pocketbook.
(Center of Military History.)

Daniel Tompkins, governor of New York from 1807 to 1815, was an implacable enemy of the Erie Canal who claimed, "There's not a drop of aristocratical or oligarchial blood in my veins."
(*Daniel D. Tompkins* by John Wesley Jarvis. New-York Historical Society, accession #1954.213.)

Martin Van Buren, one of New York's ablest politicians, was an early opponent of the Erie Canal, but then, reading the political winds, became an ardent supporter.
(Chicago Historical Society.)

James Geddes.
(Erie Canal Museum, Syracuse, New York.)

Benjamin Wright.

Canvass White.
(from *History of Saratoga County, New York, with Illustrations and Biographical Sketches of Its Prominent Men and Pioneers* by Nathaniel Bartlett Sylvester.)

Nathan Roberts.
(Erie Canal Museum, Syracuse, New York.)

There was not a single civil engineer in the United States when construction on the canal began in 1817. The four chief builders and technological innovators of the Erie Canal had previous experience only as surveyors. When completed in 1825, the canal operated exactly as planned.

The Genesee River at Rochester is so wide that the aqueduct carrying the canal across it was more than three city blocks in length. This was one of the largest structures on the Erie Canal.

(*First Erie Canal Aqueduct over the Genesee River plan* by Everard Peck, from the Rochester Historical Society Publication Fund Series, Vol. XI, p 306.)

Clearing the thick forests along the canal's route was a challenge. To pull up the stumps, the crews devised a giant contraption on two huge wheels with a smaller wheel on the axle. When a team of horses pulled a rope coiled around the inner wheel, a chain around the stump tightened and yanked the stump free.

(Drawing by Anthony Ravielli. American Heritage Picture Collection.)

THE LOCK.

The Erie Canal required eighty-three locks to raise or lower boats a total of 675 feet over the 262-mile route between Albany and Buffalo. The locks, dug with little more than shovels, blasting powder, and human sweat, were 90 feet long and 15 feet wide. Most locks moved boats up and down a distance of 8 feet 4 inches.

(From *Marco Paul's Travels on the Erie Canal* by Jacob Abbott, 1852. The New York State Library, Manuscripts and Special Collections.)

"Low Bridge!" Packet boats were pulled by three horses or mules, with a roof providing fresh air and sightseeing. But low bridges connecting lands divided by the canal could push inattentive passengers overboard while taking the air.
(The New York State Library, Manuscripts and Special Collections.)

Just twenty-five miles north of Lake Erie, the canal had to scale the precipitous escarpment over which Niagara Falls spills its torrents a short distance to the west. Five oversized locks climbed sixty feet up the Cliffside. Then, with only crude derricks, shovels, blasting powder, and sweat, the work crews slashed seven miles through the solid rock face to carry the canal and its towpath southward toward Buffalo.
("Excavating the Deep Cut West of Lockport," 1825. Canal Society of New York State.)

A fantastical monument to the visionaries of the Erie Canal.

A celebration poster for the opening of the canal.
(The New York State Library, Manuscripts and Special Collections.)

Fireworks over City Hall on the night of the ceremonies of the Wedding of the Waters.

and fifty miles were about half completed. Work on the eastern section had begun. A year later, the work on fifty miles of the western section had been finished; boats were also traveling back and forth between Utica and Little Falls on the eastern section. Then things began moving even faster. By the time Clinton was out of office a year, people could sail on the canal from a point west of Rochester all the way to Albany and into the Hudson River. Now New York City lay an easy sail of 150 miles down the river.

CHAPTER 14

UNWEARIED ZEAL

The contrast between the divisive behavior of the politicians and the extraordinary achievements of the engineers is the most remarkable feature of the whole story of the Erie Canal. While De Witt Clinton and his opponents were knee-deep in political battles, construction on the canal moved merrily along. It was as though Albany had no reason to exist except as a city that happened to be located at the junction between the canal and the Hudson.

America had no trained civil engineers when construction began. No one could have predicted how men with such little experience could overcome the succession of obstacles lying in their path, especially as they had no canal of comparable size and complexity as a model to guide them. They were going to confront broad rivers, chasms, fissures, precipitous cliffs and thunderous waterfalls, deep and narrow valleys, squishy sandstone on which to base great piers and supports, and through all of it the dense wilderness, generously populated by wild beasts and barely populated with human beings.

Perhaps the philosophical motivation embedded in the entire project drove the engineers to high success. Here nature was less an enemy to be subdued than a gift from God to be joined in the larger struggle of building a great nation. The wilderness, the waterfalls, the deep valleys were proof of the country's grandeur, which the canal would neither destroy nor remodel. As the engineers perceived it, the canal would only enhance the splendor and dignity of God's gift.

A visit to the canal today is a spiritual experience, even though much of the old canal is in ruins or has disappeared. The marvelous blend of engineering and respect for nature along the route could have been achieved only with rare imagination in the planning and astounding skill in the execution. In an era when technological change was accelerating at an astonishing pace, this beautiful work by amateurs would become one of the wonders of the world in the first half of the nineteenth century.

The leaders of this group of engineers—Benjamin Wright and James Geddes—had been judges and surveyors before they took on the task of engineering. Many years later, James Geddes's son George recalled the stubborn skepticism of the Albany legislators, who in 1816, a year before final passage of the canal bill, "tauntingly asked, 'Who is this James Geddes, and who is this Benjamin Wright . . . what canals have they ever constructed? What great public works have they accomplished?' "[1] Canvass White had studied the English canals with care but had no hands-on experience of his own. Another of the group, John Jervis, was an axeman on the canal when construction started and taught himself surveying at night; in 1820, he was given supervision of the middle section. Jervis went on to a brilliant engineering career in both canals and railroads. These were men who would dare to experiment and keep on experimenting until they got it right. Some of the most ingenious inventions came from people with the least training. For example, Jervis reported with pride that "the plan for a timber trunk for [supporting] the aqueducts [across rivers and valleys] was prepared and submitted by a carpenter, Mr. Cady, of Chittenango."[2]

The canal commissioners broadcast repeated boasts about the feats of the engineers, and Charles Glidden Haines could proclaim in 1821,

"For accuracy, dispatch, and science, we can now present a corps of engineers equal to any in the world."[3] Noble Whitford, the early-twentieth-century historian of the canal, offers a more thoughtful but typically American viewpoint. Like the architects of the nation itself—Washington, Hamilton, and Jefferson—"these men were working for a cause, for the development of their native land, and not for personal gain and aggrandizement. . . . What they did not understand they conquered by diligent study, unwearied zeal, and sound common sense."[4]

* * *

The middle section was finished during the course of 1820. Continuous transportation was now available to shipping and travel over ninety miles through flat country, with no locks, from Seneca Falls at the western end through to Rome and onward into the eastern section as far as Utica. Business boomed immediately, as traffic was fed into this open portion from both ends. Indeed, the instant success of this work strengthened the case of the advocates of the canal to complete the canal all the way from the Hudson to Lake Erie without delay, despite the obstructions being raised by a determined opposition.

The actual cost of the middle section had exceeded the original estimate of $1,021,851 by only 10 percent.[5] The accuracy of the initial projections is remarkable in view of the novelty of the task and the challenges of clearing and then digging through the virgin forests. Most of the modest cost overrun was due to changes in plans as the work progressed: for instance, stone was substituted for wood in the aqueducts, and bearing piles and weirs* had been added. Some areas provided unexpected problems in excavation, which involved extra payments to the contractors. Most important, Canvass White's indispensable waterproof lime was more expensive than the crude material provided for in the original planning.

*A weir is a channel designed to take off excess flows of water from the main waterway.

The opening of the middle section and the Rome-Utica portion of the eastern section provided the pretext for the sort of elaborate ceremony that was becoming a habit as the canal progressed, with July 4 as the favorite date. On this occasion, the celebration took place at Syracuse. The town was a burst of color on that July 4 of 1820, as a banner-waving crowd of men and women arrived aboard seventy-three gaily decorated boats and a much larger contingent traveled to the gala by land.

The commissioners' report for 1820 could not resist waxing eloquent over what had been accomplished on the canal:

> The novelty of seeing large boats drawn by horses, upon waters . . . through cultivated fields, forests, swamps, over ravines, creeks and morasses, and from one elevation to another, by means of ample beautiful and substantial locks, has been eminently exhilerating [*sic*]. The precision of the levels, the solidity of the banks, the regularity of the curves, the symmetry of the numerous and massive stone works, the depth of the excavation in some places, the extent of the embankments in others, and the impression produced everywhere along the line, by the visible effects of immense labor, have uniformly afforded gratification mingled with surprise.[6]

Toward the end of 1820, construction crews started work on the stretch running from Utica to Albany and the terminus of the canal at the Hudson River. A rapid rate of economic development followed the completion of each stretch of the canal as it flowed through charming little towns that signified their newfound prosperity with an increasing number of proud Greek revival homes. Albany itself was building a harbor facility with a wharf three-quarters of a mile long and a hundred feet wide, fully equipped with large numbers of storehouses of varying shapes and sizes.[7]

Utica was a remarkable little city, which had caught Clinton's fancy on the expedition of 1810, when he admired the elegance of the houses and the shops "well replenished with merchandise." With the canal running right through the middle of town, Utica's commercial activity blossomed despite the absence of the dramatic waterfalls and rapids that powered the textile and grist mills to the east. The local businessmen made up for this shortcoming with a wide assortment of light manufacturing and a lifestyle that attracted a variety of cultural institutions. A traveler in 1829 reported the town was "a really beautiful place," with wide streets running at right angles to one another and a main street, State Street, that is "in no respect inferior to [Broadway] in New York."[8] (A twenty-first-century traveler to Utica would agree. Even Union Station for the railroad line is an awe-inspiring sight in the modern world.)[9]

Forty miles east of Utica, Canajoharie was known for its very popular footraces in the early days of the Erie Canal, but the canal's transportation facilities would convert the town into a large manufacturing center based on the headquarters of the Beech-Nut Company, a chewing-gum and candy firm. Farther along the Mohawk valley, there is a village originally known as Remington's Corners in honor of Eliphalet Remington, the inventor of both the Remington rifle and the Remington typewriter. In 1828, Remington took advantage of the town's location on the canal to develop global markets for his remarkable and highly successful inventions, but in 1843 Remington decided he did not like having his hometown named after him, and insisted on a change. The postmaster—perhaps inspired by the town of Troy farther to the east—proposed Ilion, the Greek word for Troy, a suggestion immediately approved by the local citizens as well as Remington.

By the end of 1821, navigation on the eastern section was open on the stretch of twenty-four miles from Utica to Little Falls, and the difficult excavation from Little Falls to Albany was well advanced. Furthermore, money was now coming in as well as going out. The canal collected tolls on everything from salt, gypsum, grains, timber, and bricks to passenger boats charging 5¢ per mile. Toll revenues in 1821 were nearly five times the 1820 revenue of $23,001, which covered 13 percent of the annual interest payments due on the $2.9 million canal loans outstanding at year's end. And the work had only just begun.

The work to the east of Utica began with a particularly difficult stretch where the land drops off sharply by 105 feet over about eight miles, requiring thirteen locks one after another. Beyond that point, construction of the canal along the eighty-six miles from Little Falls to the Hudson River was a tough test of the skills of the engineers, through some of the most spectacular scenery along the entire route—scenery described neatly by one traveler as "too sublime for my dull pen."[10] The valley is seldom straight, the hills on both banks have many steep slopes right down to the river, the course of the river is narrow, and the level of the land slants steadily downward. The narrow gorge at Little Falls is the point where the Mohawk slices through the mountains. The drop from Schenectady to Cohoes Falls near Troy is the steepest part of the canal. And at the end of the line, there was the seventy-foot cataract of Cohoes Falls itself to contend with. Benjamin Wright and Canvass White spent many hours surveying all the surrounding areas in search of a more congenial route to bypass the Mohawk valley, but finally had to settle on following the river right down to the Hudson.

The landscape at Little Falls was the most defiant in the entire valley. Although there would have been no Erie Canal without that preglacial gorge at Little Falls—described by a traveler in 1829 as "the wildest place on the canal"—there nearly was no Erie Canal at all because of it.[11] Squeezed between steep banks rising as high as five hundred feet above the level of the river, the torrent of white water pours down over a broad tumble of rocks and then falls over forty feet in less than half a mile through the narrowest stretch of the entire canal from Buffalo to the Hudson. The village of Little Falls connects with the canal along a large aqueduct about thirty feet over the rushing waters. Today, Little Falls combines its wild scenery and sparse remnants of the original remarkable set of locks with a cluster of high quality art dealers and a fine museum.

In June 1822, De Witt Clinton passed through this country on canal business and stopped at a small town called Palatine Bridge, twenty miles east of Little Falls, and, in a typical comment, observed that "the operations on the Canal exhibited a scene of bustle and business. . . . The country was filled with crops of pease, rye, wheat and Indian corn. . . . We saw the king bird . . . boblinkus . . . crow, blackbird, robin . . . and the kingfisher. . . . [A] superb lilac in full blossom was remarkable on account of its size and beauty."[12]

By November 1821, just before activity closed down for the winter, the engineers began to fill the canal with water through the grueling sixty-two-mile stretch from Little Falls to Schenectady. The work had been well done: no meaningful leaks or breakdowns in the canal walls developed. But most of the Mohawk's passage between Little Falls and Schenectady tended to be rocky, uneven, and roiled by rapids. The engineers would have preferred to use the river if at all possible, but the effort to make the river navigable here had led the old Western Inland Lock Navigation Company to sorrow and ruin among shallows, white water, floods, and falls. In some areas, the passage was so narrow the engineers had to blast a route out of the rocks along the sides of the banks—including space for the towpath as well as the canal itself. Benjamin Wright was concerned they might even have to tunnel right through the cliffs in some places. In other areas, they built the canal high above the water, along embankments strong enough to be safe from the Mohawk's violent seasonal flooding. Thirteen locks with a total drop of ninety feet were necessary between Little Falls and Schenectady.

The work became even more demanding below Schenectady, where the land drops over two hundred feet in the sixteen miles to Troy. Planning the construction of twenty-seven locks—almost one-third of the canal's total—over this short stretch was just part of the problem. Traveling toward the Hudson from that point, the canal ran along the south bank of the Mohawk, but the shape of the hillsides on that bank was steeper and more irregular than on the north side. Crossing back and forth from one bank to the other seemed like an awkward solution, but the engineers figured they could actually save money if they bypassed the hurdles on the south side. Two substantial aqueducts were needed, one to go from south to north and one to bring the canal back to the south side. The northbound aqueduct just below Schenectady was 748 feet long, supported by sixteen piers that allowed the canal boats to sail along 30 feet above the roaring river. The other aqueduct, just above Cohoes Falls, a cataract as impressive for its width and thick white foams as for its height, brought the waterway back to the south bank on an even more imposing structure, 1188 feet long with twenty-six piers supporting it. This was the longest of the eighteen aqueducts on the canal. When the canal arrived at Cohoes Falls, the crews had to chop the route out of the sheer rock walls at a great elevation above the river—a job completed in eighty days, even though the English engineer William Weston had predicted it would require two years.[13]

The crisscross of the Mohawk, executed in large part in 1823–1824, saved $75,000 in construction costs, or nearly 30 percent of what it would have cost to go all the way on the south side of the river. Total outlays for the Schenectady-Albany section came to $540,000, most of which was spent on the sixteen-mile stretch from Schenectady to Troy. The cost per mile was thus more than double the per-mile cost of $12,000 for the portion of the canal from the Seneca River to Utica.

Later, in 1824, when the whole complicated and treacherous construction from Utica to the Hudson was complete, with all the daunting obstacles overcome, the commissioners could hardly believe the extraordinary achievement. Their report of that year admitted that if construction had begun there, without the experience gained from the earlier work on the relatively easy middle section, building the canal would have been "entirely abortive," and the completion of the project would have been postponed for as long as a hundred years. At that happy moment, they expected the full canal to be open and operational within the next twelve months.

* * *

Although all twenty-seven locks between Schenectady and Albany were kept open day and night along this short stretch, the frequent tedious waits for canal boats moving in both directions limited the traffic pretty much to heavy cargo. The trip required twenty-four hours even when the traffic was light and much longer when it was crowded.

Passengers had the alternative of a three-hour stagecoach ride from Albany to Schenectady (the fare was 62¢ in 1829).[14] This choice gave westbound travelers the opportunity to leave Albany late in the afternoon, enjoy dinner and a comfortable sleep in Schenectady, and then board the packets, or passenger boats, early the next morning. Clinton may have found Schenectady a dull town, "destitute of books," but its strategic location between the steeply sloping landscape toward Albany and the long span of flatland toward Utica made it a busy place indeed for both passenger stopovers and transshipments of cargo.

The road from Albany to Schenectady was not a straight line. In order to travel where the hills were less steep, it ran slightly northeast to Troy on the Hudson before heading west again to Schenectady. The canal followed the same route. Troy—a town too plain to fuss with Greek words like Ilion—earned its place in history when a local woman named Hannah Montague, in a star example of American ingenuity, decided one day to cut off the collars and cuffs of her husband's shirts and wash them separately. And thus began the nineteenth-century institution of detachable collars and cuffs. That was just the beginning of Mrs. Montague's contribution to mankind. In the 1830s, Troy became the home of the Cluett, Peabody shirt company, manufacturer of collars and cuffs and shirts, and soon to become famous for the Arrow shirt that graced billboards and printed advertisements with many handsome faces for the next 150 years.[15]

Nearby, the Quakers were the first whites to settle in the small town with the romantic Indian name of Niskayuna. The Indians in Niskayuna cultivated a strain of corn with strong stalks but no pods—a perfect material for brooms. The Quakers duly took note and employed the canal to launch what later developed into a large-scale industry supplying the whole world with their brooms.

* * *

The Great Western Turnpike ran along the proposed route of the western section of the canal, beyond the Seneca River toward Rochester and ultimately Lake Erie. This roadway would be no competition to the Erie Canal. Originally an Iroquois trading path and now New York Route 5, the turnpike was so narrow wagons had difficulty passing one another, and in the spring they often got stuck in such deep mud that they could not move. The largest towns along the turnpike were barely 6000 people when canal construction began.

The hurdles to be overcome in the western section of the canal would turn out to be every bit as intimidating as in the east. There were violent undulations in the land across the canal's route as it approached Rochester. The Genesee River in Rochester itself, running south-north, was the most extensive river crossing in the whole system, and an awesome confrontation with the Niagara escarpment would have to be resolved before the easy sail to Lake Erie would come into view.

Just east of Rochester, where in 1808 James Geddes had so joyously cried, "Eureka!," the Irondequoit Creek runs northward through a deep and narrow valley on its way to Lake Ontario, cutting right across the east-west route of the canal. The waterway was misnamed: it was a lot more than a creek, as boats up to thirty tons made use of it on their way to Lake Ontario and beyond to Canada and the St. Lawrence River.

It was no simple matter to get a forty-foot-wide ditch full of water across a U-shaped chasm with about one mile between the top of one side and the top of the other. The low point of the U varied from forty to seventy feet below the upper rims. Consequently, the slopes down to the creek were too steep and the valley itself too narrow to provide room for locks down one side and then back up on the other side. How, then, could the crossing be established?

Geddes had originally visualized a huge embankment to carry the canal from one side of the valley to the other, where the creek would flow underneath while "boats would one day pass along on the tops of the fantastic ridges, [and] posterity would see and enjoy the sublime spectacle." When the engineers started actually designing this embankment, they found the soil in the area was too porous and crumbly to hold together under the weight and motion of the canal. They decided to replace Geddes's scheme with a long wooden aqueduct that would carry the canal at the height of a five- to six-story building from one edge of the gap to the other. But before they had signed all the contracts, the engineers had second thoughts about this job: the winds blowing in from nearby Lake Ontario could easily topple so tall and narrow a structure, top-heavy with the canal's water.

The only solution was to return to Geddes's original proposal: fill in the valley right up to its rim with a massive embankment to create a passageway that would carry the canal from one tip of the U to the other. In order to avoid blocking Irondequoit Creek in the process, a space had to be opened at the bottom of this huge pile of earth to allow the creek's northward flow of water to wend its way up to Lake Ontario. This stone culvert, or transverse drain, was twenty-five feet high, thirty feet wide, and a hundred feet long. It had to be carefully arched to support the great bulk of the embankment looming another forty feet above it.

Imposing size was not the only problem presented by the embankment. As the earth at the trough of the valley, on either side of the creek, consisted largely of quicksand, the gigantic mound could easily slide away in a mess of mush and carry the canal along with it. To eliminate the risk of such a catastrophe, the engineers had to sink over nine hundred twenty-foot log pilings deep into the ground below the levels of the quicksand. The local earth was not cohesive enough to hold together even with the pilings, requiring the importation of different kinds of dirt from the surrounding country.*

The building of the Irondequoit embankment was where the Irish immigrants—as many as three thousand of them—made one of their largest contributions to the canal. The Irish were soon famous for their colorful ways and lilting songs such as:

> We are cutting a Ditch through the gravel,
> Through the gravel across the state, by heck!
> We are cutting the Ditch through the gravel,
> So the people and freight can travel,
> Can travel across the state, by heck![16]

J. J. McShane, the contractor who hired the largest number of them, was formerly a prizefighter and had worked on canals in Ireland. McShane was a tough taskmaster, but he did provide a jigger boss, a young man who went along the line handing out tots of whiskey several times a day. Even with the whiskey, working conditions must have been horrendous. The colossal mass of earth and its culvert had to be assembled with construction equipment little different from what men had used for centuries: horses and wagons, oxen, wheelbarrows (including horse-drawn three wheelers for the heaviest material), shovels—and bare hands. The crews toiled all day, even in the hottest months of the summer, before finishing the job in October 1822, after about two years of effort. The work frequently continued into the night, with nothing more than bonfires to light the canal's progress.

Irondequoit led Philip Freneau, the first American poet of note, to declare the canal "a work from Nature's *chaos* won. . . . Ye artists . . . proceed!—and in your bold career may every plan as wise appear."[17] The design, the workmanship, and the hard labor all paid off without a hitch.

*Richardson's Canal House, an inn dating back to 1818 in Bushnells Basic, the town at the eastern end of the embankment, is the oldest original Erie Canal inn still in place and in much of its original form.

A well-timed rainstorm filled the canal right after the completion of the embankment, and boats immediately began traveling the waterway eastward all the way to Little Falls. Before long, these eastbound boats were passing boats in the opposite direction on their way to Rochester. One of the first of these vessels arrived at Rochester from Utica, 152 miles away, carrying eight families—a total of sixty people at a cost of $1.50 each.[18] The Marquis de Lafayette himself came through in 1824, amazed as the canal "pursued an aerial route."[19]

Rochester—the future home of George Eastman, the Eastman Kodak Company, and the Xerox Corporation—was headed for spectacular growth and economic importance from the very beginning. John Howison, a traveler passing through in 1820, marveled at the population of over 3000 people for a town less than ten years old. He also noted the "spacious streets . . . well-furnished shops, and the bustle which continually pervades them." But the first steps to becoming a boom town this early in the Industrial Revolution in the United States began to transform natural beauties into ugly industrial sites. Howison noted that the magnificent Genesee Falls, which provide Rochester with "fine water power . . . are unfortunately surrounded with machinery; for the rattling of mills, and the smoke of iron founderies [*sic*] . . . neither harmonize well with the wildness of uncultivated nature, nor give any additional interest to a scene where they are so manifestly out of place."[20]

By 1827, Rochester's population was approaching 8000 and, as Geddes had predicted fourteen years earlier, the embankment became a great sightseeing attraction where people had to crane their necks to see the canal boats sailing along so far above them. Twenty-five years later, at 36,000, there would be almost as many people living in Rochester as in Buffalo.

* * *

At Rochester, the canal had to cross over the Genesee River at a point where the current is unusually swift—the Genesee Falls are right next to the canal route. The Genesee was so wide that the aqueduct carrying the canal across it was more than three city blocks in length, supported by nine Roman-style stone arches, each fifty feet across, and one smaller arch at each end. This structure was the largest on the entire canal.

The chief engineer on the Genesee aqueduct, William Britton, had just finished building a new state prison at Auburn when work began in the fall of 1821. In the interest of increasing his own profit margins, Britton brought along twenty-eight convicts to cut and shape the stones, and then arranged for additional groups of convicts to join in a short time later. This scheme turned out to be an unhappy arrangement for the canal, because the convicts needed guarding and some managed to escape. To make matters worse, a heavy flow of ice during the first winter carried away the foundations for one of the piers. Britton conveniently died at the same moment, resulting in a new contract and a fresh start the following spring.

Britton's passing may have been a relief to all concerned, but there were continuing problems with both raw materials and labor force. The local red sandstone from the nearby town of Medina proved to be unsatisfactory for sustaining the great weight of the bridge on the rock bed of the Genesee—although over time Medina sandstone was used in buildings all over America. Medina would also compensate for the shortcomings of its sandstone in serving the canal with what the citizens claimed to be "the most pretentious hostelry in the region," the Eagle Tavern, right on the canal and topped with a gilded bird on its belfry.[21] The engineers had to strengthen the sandstone with limestone brought from the area of Lake Cayuga, over fifty miles away. In addition, the local population was too small to provide enough workers, so once again Irish immigrants were brought in to do the hard labor. After a winter marked by unusually heavy snows and torrential rains at the end of summer, the Genesee aqueduct was finally completed in September 1823, eleven months behind schedule and $83,000 over the original cost estimates.

The engineers and their crews had created a miracle—boats crossing a wide river 450 miles from the sea through what had once been a dark wilderness.[22] The aqueduct they built was so strong it could support two thousand tons of canal water. No disappointments could stop the gala celebrations when the job was done.

The workmen began the fun on September 10, as soon as they received their final payments from the contractors. Four weeks later, a cluster of colorful boats gathered, the bands played, the orators orated, the clergymen sermonized, and the highest-ranking notables enjoyed a lavish feast at a local tavern, washed down with twenty-five toasts, including one "to the ladies of Rochester—the brightest ornament of our day's celebration." And this about a city that had been described in 1809 as "a Godforsaken place inhabited by muskrats, visited only by straggling trappers, through which neither man nor beast could gallop without fear of starvation or ague."[23]

The most dramatic moment of these celebrations occurred when a canal boat arrived to deliver fresh oysters to Rochester, over four hundred miles from the Atlantic Ocean. The oysters had traveled so rapidly along the canal they were still fresh enough to be eaten even at this distant outpost of civilization on the western frontier.

While the oysters were the most exciting product to arrive in Rochester at that moment, incoming boats also brought in gallons and gallons of beer and whiskey. More important, 10,000 barrels of flour departed eastward from Rochester in the first ten days of operation, followed by large quantities of pork and a wide variety of forest products. In just the first year, canal authorities at Rochester collected $9802 in tolls.[24]

The 10,000 barrels of flour were a prediction of things to come. With the great power of Genesee Falls to keep the mills turning, Rochester was soon known as the nation's Flour City. The citizens of Rochester preferred the city's more auspicious title, "Young Lion of the West." The English painter William Henry Bartlett caught the entrepreneurial spirit of Rochester's booming prosperity when he observed that the city rested on the edge of "the only instance in the known world of a cataract turned, without the loss of a drop, through the pockets of speculators."[25] But the American painter Thomas Cole took a less cynical view: "I should consider Rochester one of the wonders of the world—there is a large and handsome town that has risen in the midst of a wilderness almost with the rapidity of thought. Future ages shall tell the tale of the enterprise and industry of the present generation."[26] And De Witt Clinton was typically enthusiastic when he arrived in Rochester in June 1823 and recalled the dark wilderness along the river when he first came through in 1810. The bridge was "a sublime work." The town itself captivated him: "A growing place. . . . The hum of men—the bustle of business—and streets crowded with building stone."[27]

Unfortunately, within a decade the sublime work over the Genesee would become a memorial of the past. The structure was so narrow that boats coming in opposite directions could not pass each other, leading to endless squabbles and often fistfights over which side could claim right of way. The bridge also leaked badly, with precious canal water dripping day and night down into the Genesee below. In 1833, a new bridge replaced the old.

* * *

On September 10, 1823, Clinton's diary notes that he "went to visit the first passage of sloops through the lock and dam at Troy. . . . A splendid day." On October 1—at almost the same moment as the celebrations in Rochester—the engineers declared the eastern section complete; the Erie Canal was now open to uninterrupted travel all the way from the Genesee River to Albany—a distance of over 250 miles. In addition, the canal commissioners could pronounce the north-south Champlain Canal fully open for business. This canal covered 64 miles through twelve locks from Whitehall, a little village squeezed into a narrow valley at the southern shore of Lake Champlain, to Watervliet, located just seven miles north of Albany.

The first operating federal arsenal in the United States had been established at Watervliet in 1813. (In the twentieth century, this arsenal would be the sole manufacturing facility for large-caliber cannon, including the navy's giant sixteen-inch battleship guns.) As an important junction between the two canals, Watervliet was equipped with a weigh station and was also a center for paying canal boat operators, which accounted for its fame as a center of saloons, gambling, and prostitution.[28]

There were festivities at Albany that October 1 to celebrate the simultaneous opening of the east-west and north-south linkages, including military performances and a band from the cadets at West Point, the usual roaring cannon and expressive speeches, as well as an elaborate aquatic procession. Cadwallader D. Colden claimed, "The pencil could not do justice to the scene presented on the fine autumnal morning when the Albany lock was first opened." The crowds filled the windows and the tops of the houses, jammed the open spaces in the fields, and lined the banks of the canals for a number of miles.[29]

In the place of honor at the head of the procession was a handsomely decorated packet boat named the *De Witt Clinton*, with the great man himself aboard. At the climactic moment, according to Colden,

> The waters of the west, and of the ocean, were then mingled by Doctor Mitchill, who pronounced an epithalamium upon the union of the River and the Lakes, after which the . . . gates [of the last lock] were opened, and the *De Witt Clinton* majestically sunk upon the bosom of the Hudson. . . . Numerous steam boats and river vessels, splendidly dressed, decorated the beautiful amphitheatre formed by the hills which border the valley of the Hudson, at this place; the river winding its bright stream far from the north, and losing itself in the dis-

> tance to the south;—the islands it embraced;—the woods, variegated by the approach of winter, a beauty peculiar to our climate;—the wreathed arches, and other embellishments, which had been erected for the occasion, were all objects of admiration.[30]

Then, among the usual long series of speeches, Clinton delivered the keynote address on the wonders the canal had achieved. The following evening, he and his wife led the cotillion at the Albany Canal Ball. All of these splendid arrangements were no coincidence: Clinton himself had orchestrated the entire performance.[31]

As events would play out in the months ahead, the bright drama of these ceremonies would be precious moments for Clinton. Unrelenting hostility in the legislature would lead to the deepest disappointments and the darkest moments of his entire career. Fully aware of the growing risks to his political future, his friends hoped to capitalize on the opening of the canal into the Hudson River by purchasing the *De Witt Clinton* as an icon they would display in the next gubernatorial electoral campaign, after which they planned to plant it in cement in New York City as a public monument.[32]

* * *

With the completion of the opening to the Hudson, the way was now clear for two-way water transportation between the Atlantic Ocean and Rochester, a span of some 375 miles. The engineers could take pride in the great feats they had accomplished on the steep stretch from Albany to Schenectady, past the roaring rapids of Little Falls, over the deep embankment at Irondequoit, and finally across the broad Genesee River.

But the final section of the canal running from Rochester to the terminus on Lake Erie imposed its own set of challenges. Even before work could begin, the route itself had to be determined, and that was no simple matter. There was no way the canal could bypass or circumvent the huge Niagara escarpment blocking the passage between Rochester and Lake Erie. Somehow, canal boats were going to have to climb up the escarpment if they wanted to go all the way to the Great Lakes. And after all that had been accomplished, some of the most bitter political wrangling over the canal still lay in wait.

PART IV

The Stupendous Path

CHAPTER 15

A NOBLE WORK

The most direct course from Rochester to Lake Erie was a southwestern diagonal ending up in the vicinity of Buffalo. This choice had the additional attraction of passing through Batavia, the local headquarters of Holland Land territory, where Joseph Ellicott and his property owners would welcome the additional trade from the canal as a nice payoff for their donation of 100,000 acres to help defray the cost of building the canal.

Unfortunately, the summit of this route was seventy-five feet above the level of Lake Erie, which blocked the lake from serving as a source of water for the canal. The local streams were much too small to serve the purpose. The diagonal course through Holland Land Company territory would not have brought much satisfaction to Ellicott or anyone else if the ditch ended up with no water running through it. Gouverneur Morris would have been amused. Others may have considered his vision of "tapping Lake Erie . . . and leading its waters in an artificial river, directly across the country to the Hudson River" to be a romantic dream, but here tapping Lake Erie's waters was essential.

There was an alternative route, which would run due west from Rochester, staying close to the shores of Lake Ontario for seventy miles;

at that point, the canal would make a ninety-degree turn southward and head on down to Lake Erie, thirty miles away. Although this route was longer than the diagonal through Holland Land country, it would carry the canal without any locks through flatlands between Rochester and the turn to the south. Between the turn and Buffalo, there was plenty of water, most of it coming from Lake Erie and the nearby Tonawanda Creek.

But the turn itself would be the most difficult and the most spectacular feature of the entire canal. Immediately after curving to the south, the canal came face-to-face with the steep and forbidding cliffs of the precipitous escarpment over which, seventeen miles to the west, Niagara Falls spills its torrents. There was no way to avoid this nearly vertical climb of over seventy feet. Even worse, the geological structure consisted almost entirely of solid rock. Yet this had to be the choice because of the inadequate water supply situation on the more direct route.

The commissioners invited several engineers to submit plans for this intimidating project. The proposal they accepted was presented by Nathan Roberts, a staff engineer; his dramatic but effective design, although modernized, functions today precisely as he conceived it nearly two hundred years ago. Roberts was forty-six years old in 1822 (which meant he was a child of 1776), a native of Canastota, New York, who had spent his younger years both speculating in land and teaching mathematics in the local schools. The combination turned him into a master surveyor, and he had recently served with much success as chief engineer on the stretch of canal between Rome and Syracuse.

The scheme was to begin by carving a flight of locks out of the escarpment, rising one immediately above the other like a giant staircase. "Flights of locks" were nothing new, but the magnitude of Roberts's design was much greater than anything attempted before. The canal would have to climb about sixty feet up the embankment. In order to reduce the number of locks necessary to achieve this daunting objective, Roberts decided to make his locks twelve feet high instead of the eight feet four inches of all the other locks on the canal. Then he built two adjacent sets of five locks instead of just one set. With these "double-combined locks . . . working side by side," traffic could move in both directions, up and down, at the same time, preventing bottlenecks when boats headed one way would meet up with those coming the other way.[1]

The experience was like a slanting fresh-air escalator ride up or down the equivalent of six stories in a modern building. A merchant who traveled on the very first boat climbing up the escarpment on these locks was bowled over by the experience: "I was more astonished than I ever was by anything I had before witnessed," he wrote.[2]

But Roberts's five locks did not carry the canal all the way up to the highest point of the escarpment, which still loomed another ten feet and more above the level of the fifth lock. This was no oversight. Surveys by Roberts and James Geddes had shown that the escarpment tilted gradually downward from that point. Instead of having to create at least one and perhaps two more locks to lift traffic to the very top of the cliffs, Roberts now slashed right into the solid rock face to sculpt a channel carrying the canal, and a towpath beside it, for seven miles in a straight line southward to the town of Pendleton, where the slope of the escarpment had declined to the same level as Roberts's channel. At Pendleton—whose nearby hardwood forest would become a world-famous source for ship masts and barrel staves—the canal connected to Tonawanda Creek, which flows in a gentle fashion southwesterly and empties into the Niagara River, right next to Lake Erie about ten miles north of Buffalo.

* * *

The work on this massive project began at its north end at a small settlement of 3 families, situated just eighteen miles south of Lake Ontario. Soon to be appropriately named Lockport, the town had increased to 337 families by January 1823, when the contracts had been let and the first digging had begun. Lockport's population would reach 3000 people by the time the job was done two and a half years later, not counting nearly 2000 laborers working on the construction of the locks and digging the channel—the Deep Cut—through the escarpment. Many of these men were Irish who would remain in Lockport and give it a distinct Irish flavor for a long time to come.

Lockport became a major producer of electric power generated by surplus waters from the canal. The town was also a prime example of how rapidly the Erie Canal would stimulate economic development on a wider scale and attack the environment in the process. A vivid description of what "improvement" meant at Lockport was provided by Frances Trollope, the cantankerous mother of the famous English novelist Anthony Trollope, who came to America in 1828, just six years after Lockport came into being and only three years after the completion of the canal. Having traveled from New Orleans up the Mississippi and Ohio rivers to Cincinnati, Mrs. Trollope was now on her way to New York City. She deplored nearly everything she found here, including her trip on the Erie Canal—which was the route of choice to take her to New York—but she was especially vociferous about Lockport:

> Lockport is, beyond all comparison, the strangest looking place I ever beheld. As fast as half a dozen trees were cut down, a factory was raised up; stumps still contest the ground with pillars, and porticoes are seen to struggle with rocks. It looks as if the demon of machinery, having invaded the peaceful realms of nature, had fixed on Lockport as the battle ground on which they should strive for mastery. Nature is fairly routed, and driven from the field, and the rattling, crackling, hissing, splitting demon has taken possession of Lockport for ever. We slept there, dismally enough. I never felt more out of humour at what the Americans call improvement; it is, in truth, as it now stands, a most hideous place, and gladly did I leave it behind me.[3]

The Deep Cut began just south of Lockport at a level of about thirteen feet below the rim of the escarpment, but over the next mile and a half the summit of the escarpment rose to a point where the excavation was thirty feet deep. From there, the level gradually descended until it reached Tonawanda Creek. The wonder of the whole project was in digging the channel to such extended depths. The rocky composition of the escarpment was so solid and tough that the effort to drill holes for blasting often ended up with broken drills and no holes. Even special equipment ordered from New York and Philadelphia failed. Finally, a local man found a way to temper and harden the steel in a fashion that worked.

But creating a hole for the blasting powder was just the beginning. Once the gunpowder was deposited in the hole, and the fuse lit, the explosion itself was in most cases a violent and dangerous affair as huge pieces of hard rock went soaring and careening into the air in all directions, many of them landing on the streets of Lockport itself. Often a schoolboy was elected to light the fuse, on the theory he could run to safety faster and more nimbly than a grown man could make it over the jagged surface. Then there was the problem of removing the enormous heaps of rocks the explosions left lying at the bottom of the cut. For this task, the workers developed a kind of derrick, a boom for swinging buckets down into the cut to be filled with the shattered hunks of rock, after which horses pulled the cables that drew the debris up to the surface.[4] Contemporary drawings show a long line of these strange booms, one after another, along the sides of the channel for as far as the eye can see. Even after the channel had been excavated, the narrow towpath had to be carved out of its side, often fifteen feet or more above the level of the water, so that the animals could continue to provide motive power for the boats.

The crews broke through the last few feet of the cut at Pendleton in October 1824, and the entire undertaking was finally completed in June 1825. The usual celebrations took place, with the representatives of Masonic lodges in this instance playing a prominent role, in full regalia.[5]* While an audience of about five thousand spectators watched the proceedings, the Masons marched down to the foot of the locks and placed a bronze capstone memorial there. It read, "Let posterity be excited to perpetuate our free institutions, and to make still greater efforts than our ancestors, to promote public prosperity, by the recollection that these works of internal improvements were achieved by the *spirit and perseverance* of REPUBLICAN FREE MEN."[6]

In 1831, when the marvels of Lockport were fully operational, an English barrister with the unlikely name of Henry Tudor wrote home that "it certainly strikes the beholder with astonishment, to perceive what vast difficulties can be overcome by the pigmy arms of little mortal man, aided by science and directed by superior skill."[7] About the same time, a German tourist looked down from the top of Roberts's combined double locks and observed that this was "a noble work for so young a country."[8] The work was indeed noble.

* * *

We can appreciate the magnitude of the achievement at Lockport by comparing the Deep Cut to its counterpart in the construction of the Panama Canal some eighty years later, when George Goethals connected Atlantic waters to the Pacific by hacking the nine-mile Culebra Cut through the Cordillera Mountains. The one hundred million cubic yards of earth dug out to create the Culebra Cut was the largest excavation in history. The work took six years, blew up sixty million pounds of dynamite, and moved enough dirt to build a Great Wall of China from San Francisco to New York. The crews labored in heat that seldom fell below a hundred degrees, and rock slides were a constant and devastating obstacle to progress. Hundreds of men were killed on the job. No wonder the Culebra Cut gained the nickname of Hell's Gorge.[9]

*Thirteen signers of the Declaration of Independence were Masons, and fourteen presidents, including George Washington, have been members of the group.

In the 1820s, Nathan Roberts, an engineer of limited experience, created the seven-mile Deep Cut at Lockport in three years with nothing but shovels and wheelbarrows, some animals, crude blasting powder, fragile hand-held drills, and the bare hands and broad backs of the workers. Nothing like it had ever been attempted before. Imagine what Roberts could have accomplished with Goethals's equipment, as Theodore Roosevelt reported after an inspection of the works at Panama in 1906:

> The huge steam shovels are hard at it; scooping huge masses of rock and gravel and dirt previously loosened by the drillers and dynamite blasters, loading it on trains which take it away to some dump. . . . Little tracks are laid on the side hills, rock blasted out, and the great ninety-five ton steam shovels work up like mountain howitzers eating into and destroying the mountainside. It is an epic feat and one of immense significance.[10]

But there were significant parallels between the two projects, quite aside from digging through the mountains to make way for a canal. The driving vision of the Erie Canal was to link the Atlantic states by water with the west and make a great nation. Roosevelt, having enthusiastically promoted the Erie Barge Canal as governor of New York, envisioned the Panama Canal as the new passage between the two oceans to make the United States a great global power.[11] There was an echo of the Erie Canal in the slogan inscribed on the Tiffany and Company shield created for the Panama Canal in 1906: "The Land Divided—The World United."[12]

* * *

Dividing the land of New York in order to unite the Atlantic states with the west may have been a noble mission, but nobility was nowhere to be seen where the political squabbles surrounding the Erie Canal raged on. The most protracted wrangling over the route of the canal centered on the selection of the western terminus at Lake Erie.

Protracted is perhaps too mild a word to describe the brutal and seemingly endless controversy among competing business interests, politicians, and engineers.* The conflict began as far back as 1816 and was still raging while the Deep Cut from Lockport to Pendleton was nearing completion. Large economic rewards were at stake, for the canal's port on Lake Erie would be the equivalent of New York City's position on the east coast—the crucial gateway between the Atlantic states, the Atlantic Ocean, and the great territories, lakes, and rivers to the west.

Unlike the eastern end of the canal, where Albany and New York City were obvious choices, the topography of the shores of Lake Erie obscured the optimal solution rather than illuminating it. There were no natural harbors at any point down the lake's entire eastern shore. The closest configurations that just might serve the purpose were at two small communities, Buffalo and Black Rock, separated from each other by only three miles of dense forest at the northeast corner of the lake, almost due south of the Deep Cut at Pendleton. The nearest alternative for a port and harbor adequate to serve the purpose was Dunkirk, still in New York State but fifty miles away, on the southern shores of the lake.

Unlike the design of the entire distance of the canal from Albany to Lake Erie, the Buffalo–Black Rock contest provided the politicians and businessmen with a rich opportunity and strong incentives to interfere with the decisions of the engineers. The locals took every advantage of the opening provided to them.

Buffalo, close to the northeast corner of Lake Erie, was a small village of little importance at that time, with no more than two thousand inhabitants. Originally known as New Amsterdam, Buffalo's street plan and lakefront area had been laid out by Joseph Ellicott of the Holland Land Company, who at one time owned as much as one-third of the

*The controversy was so protracted, in fact, that, taking mercy on the reader, I have compressed it to bare bones.

property there.[13] When De Witt Clinton and the other commissioners reached Buffalo in 1810 during their survey of the canal's possibilities, they found only thirty or forty houses, a courthouse, a few stores, and a tavern where the commissioners were "indifferently accommodated in every respect." Buffalo remained as a quiet place, an attraction for tourists or a stopover for pioneers heading west, until the Erie Canal came into view.

As Ellicott's map of Buffalo reveals, the town had been built right down to the shores of the lake while at its southern end a snakelike creek wiggled lethargically into the lake. Yet Buffalo was not the obvious choice over Black Rock: it did not have anything resembling a natural harbor even though it was located on the lake. The movement of the tides offshore was strong, and the lake's waters were shallow. A sandbar prevented all but the smallest of boats from reaching land, a disability that could be cured, but only at substantial expense. This was not a place where ships could easily move in and out.

Unlike Buffalo, the little town of Black Rock was not situated on Lake Erie. It was set slightly inland, just above the mouth of the Niagara River, which flows eastward from Lake Erie into the interior and then north to the precipice at Niagara Falls. Because of its location on the river, Black Rock was a much busier settlement than Buffalo. The town had a harbor of sorts, sheltered from the winds blowing across the lake by two small islands, Bird Island and Squaw Island. The black rock after which the town was named, a one-hundred-foot outcropping from the shore, formed a natural wharf, with a landing for boats somewhat protected from the force of the current.

Most important, as Black Rock's most famous citizen, Peter Porter, was fond of reminding his colleagues on the canal commission, the Niagara River was navigable all the way up to Lake Ontario, just thirty miles to the north. The only break was at the area around Niagara Falls, where Porter's firm conveniently arranged to carry the boats around the falls and then back into the river. Before the completion of the Erie Canal, most travelers and cargo going west went by way of Lake Ontario rather than on the bumpy overland roads, with the Niagara River as their connection to Lake Erie. They kept Porter's eponymous portage operation busy and highly profitable. If the Erie Canal were to establish its terminus at Black Rock, Porter would become one of the most powerful men in the whole American economy. He would never give up trying.

But Black Rock also had problems. The current in the Niagara River flowing north from Lake Erie and past the town was too strong for sailing vessels to negotiate the short passage from Black Rock harbor into the lake. An enterprising resident, Sheldon Thompson, furnished these vessels with what he called the "horn breeze"—his team of fourteen oxen pulled the boats up the river and then delivered them into Lake Erie.

* * *

In 1816, after much discussion, the canal commissioners settled on Buffalo as the best place for the terminus of the canal. Clinton was so enthusiastic he predicted Buffalo would one day be as great and powerful as New York. But the Black Rock contingent was not about to accept this decision without a fight. They were already enjoying the prospects of business and publicity they would receive when the steamboat Peter Porter was building would be launched into the Niagara River.[14] The 338-ton *Walk-in-the-Water* made its maiden voyage in 1818, with the assistance of Thompson's "horn breeze," and traveled all the way across Lake Erie to Detroit and back.

In the same year, at Joseph Ellicott's suggestion, De Witt Clinton asked William Peacock, one of the engineers assigned to the western section, to make one final survey of the area around both Buffalo and Black Rock and come up with a firm recommendation as to which would best provide for "a safe and commodious harbor" for the canal. Ellicott had not recommended Peacock just because of his engineering abilities. Peacock had done most of the surveying on the rejected diagonal route for the canal from Rochester to Buffalo by way of Holland Land Company territory. In the process, he and Ellicott had become friends. Ellicott's interests in Buffalo had a lot riding on the decision and Peacock's friendship could make a difference.*

Peacock performed as anticipated. His detailed engineering report of January 1819 covered everything from the sand and gravel in Buffalo's harbor to the timing of the lake's currents. After recommending the construction of a pier extending a thousand feet out into the lake, Peacock concluded, in "*most decided*" fashion, that "Buffalo from its local situation is apparently the key which opens to the People of the State of New York a most stupendous path of navigation and of commerce extending the distance of more than 2000 miles."[15]

At this point, we might assume the whole matter was settled, but instead the story turns into a kind of musical comedy. In 1820 the canal commissioners reversed themselves and decided to award the prize to Black Rock. But Buffalo had no intention of giving up. An engineer named David Thomas now visited the area and persuaded the commissioners to reverse themselves yet again and anoint Buffalo as the official terminus for the canal, as they had originally decided. The citizens of Black Rock were furious, but the commissioners held fast by their decision in favor of Buffalo—at least for a while. Then, in August 1821, James Geddes and Nathan Roberts surveyed both harbors and reported back to the commissioners their judgment that Black Rock had the better facilities.

Six months later, Roberts, David Thomas, Canvass White, and Benjamin Wright went out to look over the situation one more time and recommended Buffalo. Geddes was the sole dissenter on the engineers' report and refused to sign it, but the majority prevailed. To add insult to Black Rock's injury, Porter's *Walk-in-the-Water* encountered a terrible storm and ended up a total wreck right next to the Buffalo pier. To com-

*Despite Ellicott's many successes in working for the Holland Land Company and in promoting Buffalo, his periodic depressions finally overtook him on August 19, 1826, almost a year after the opening of the Erie Canal, when he committed suicide at Bellevue Hospital in New York City.

pound the tragedy for Black Rock, the new owner rebuilt the ship in Buffalo.

In June 1822, the canal board and the five engineers met in Buffalo to make a final-final decision on the harbor and terminus. Clinton's diary reports with great detail on his trip west for this meeting, including his customary expressions of enjoyment at nature, as well as admiration of the burgeoning traffic on the canal and the enthusiastic greetings he received from the crowds along the way. He also took the time to record his complaints, as usual, which included a portrayal of the discomforts of overnight travels on the canal and his morning's reward: "Night aboard uncomfortable. Crowds of women and three children in next cabin. . . . Although like going from an oven into an ice-house, got up at night—wrapped myself up in my cloak—and stayed on deck until we passed from Oneida Creek to Rome—16 miles through gloomy swamps . . . the fog heavy. . . . Breakfasted at Rome—a good salmon."[16]

The commissioners arrived at Buffalo on the evening of June 7, after a stop to sightsee at Niagara Falls. Clinton notes that he conducted an interview that evening with a famous Seneca Indian named Red Jacket at the Buffalo courthouse before a large audience.

The next morning the commissioners began their deliberations on the choice of terminus. By unanimous vote, Buffalo was still the choice. Nevertheless, it would be February 1825 before the legislature finally ordered the commissioners "to continue and complete the Erie canal to Lake Erie at the mouth of Buffalo creek, distinct from, and independent of, the basin at Black Rock."[17]

When the legislature completed all the arrangements, Buffalo would have been unrecognizable to the commissioners who had visited there in 1810. Some of the growth was simply the result of Buffalo's location on the shores of Lake Erie and its proximity to that great tourist attraction, Niagara Falls. But by the 1820s, Buffalo was expanding in anticipation of its key role in the Erie Canal. The population of the so-called Queen City had risen to almost 2500, there were five churches, six schools, a court-house, a library, a theater, a Masonic hall, and over fifty shops of all kinds. Buffalonians could call on seventeen attorneys and nine physicians to attend their needs. They had four newspapers to keep them informed, and a constant flow of visitors and travelers heading west filled the eleven inns (quality not recorded).[18] Buffalo lived up to Peacock's promise as the key to "a most stupendous path of navigation and of commerce extending the distance of more than 2000 miles."

Not so at Black Rock. As Buffalo was beginning to flourish in its new role, the pier built by the eager citizens of Black Rock was carried away by floodwaters in May 1826. This was Peter Porter's last gasp after fighting for over fifteen years for the Lake Ontario route and against the canal's overland route from Rome westward to Lake Erie.

In 1837, Porter moved up to Niagara Falls, where he died in 1844. We can only wonder how he would have reacted nine years later, in 1853, when Black Rock disappeared as an independent locality to become part of the city of Buffalo. Today, only the black rock gives the town any identity, except for those who are familiar with its history.

* * *

While the noble work at Rochester and Lockport and the ignoble partisan struggles at Buffalo and Black Rock were going on, the rest of the canal was progressing at a rapid pace. In 1824, over three hundred bridges were built across the canal along the stretch between Utica and Albany to connect farmlands and other properties that had been split by the passage of the canal. The junction between the Tonawanda Creek and the canal was finished. New York City was booming, with three thousand new houses completed just in 1824. At Utica, Syracuse, and Troy, elaborate mechanisms to weigh boats and their cargoes were coming into operation. These ingenious devices, operating on what were known as hydrostatic locks, measured the water displaced by a boat, from which it was possible to compute the boat's weight and appropriate toll. The most handsome of these locks was at Syracuse, where the structure took the form of a classic Greek temple.

But a new outbreak of the political storms over De Witt Clinton's role in the management of the Erie Canal was gathering in Albany, a tempest that would dwarf the battle of Buffalo and threaten to delay the promised completion of the canal in 1825. The drama engulfed all the major characters in the whole long history of the Erie Canal up to that moment. No one would know until the final climax whether the ending to this drama would be a happy one or a tragedy. The dénouement had elements of both.

CHAPTER 16

THE PAGEANT OF POWER

On July 22, 1823, about six months after he had left the governor's office in a state of exhaustion and depression, purportedly ending his political aspirations, De Witt Clinton addressed the Phi Beta Kappa Society of Union College in Schenectady. His message appeared to be inspired by the ivory tower to which he had consigned himself and focused on the glories of learning. But his eloquent words reflected his true state of mind at that moment: "Pleasure is a shadow, wealth is vanity, and power a pageant; but knowledge is ecstatic in enjoyment, perennial in fame, unlimited in space and infinite in duration. . . . Its seat is the bosom of God—its voice the harmony of the world."[1]

At the age of fifty-four, the pageant of power was still driving Clinton's thoughts and dreams. In October 1823, just before the elaborate ceremonies opening the canal to the Hudson River, his diary noted a recent toast by a good friend: "D.W.C. like an old brass kettle. The harder he is rubbed, the brighter he will shine."[2] Another entry declares, "Canalling more popular than banking."[3] The diary pages that follow are full of gossip about Van Buren, John Quincy Adams (warming up for his own run at the presidency in 1824), and Representative John C. Calhoun, as well as ruminations on his own chances at the greatest prize of all. Referring to himself in the third person, Clinton cited President Monroe's secretary's comment that "Clinton would prevail; that he had seen enough to convince him." Another diary entry recounted how "when J. Q. Adams was here [he] asked what was to be done for C. His services ought to be secured to the U.S." Clinton capped these passages with a "Story of Jefferson," who was asked by his neighbors to name the greatest man in America and Jefferson responded, "D.W.C."[4] Then, "Dreams—Can we command our dreams!"[5] In November, a brief note says he heard that Van Buren had told a young lawyer in Albany "that in 10 days I would be nominated in Washington."[6]

Clinton was dreaming. His primary claim on the electorate was the Erie Canal. But by this time the Erie Canal was so popular throughout the state and so impressive in its engineering marvels that it was no longer Clinton's Ditch. All the old battles over its legitimacy were long since forgotten. Now the Erie Canal belonged to everyone, even Martin Van Buren and the Bucktails. Indeed, as we have seen, Van Buren had made a remarkable about-face from implacable opposition to unqualified support for the Erie Canal when the legislature had given its final approval in April 1817.

The Van Buren team was already hard at work on the presidential campaign of 1824, a full year in the future. De Witt Clinton was the last man they wanted to see in the job, and, being more hardheaded than Clinton, they understood how the passage of seven years had diminished the importance of his early leadership for the canal. In fact, the Bucktails now had no difficulty associating themselves with the canal and making use of it for their own purposes, such as exploiting their majority on the canal commission to dispense patronage with the sole objective of enhancing their own political support at Clinton's expense.

The fracas over the selection of Buffalo over Black Rock as the Lake Erie terminus only added to the hostility surrounding Clinton. As chairman of the canal commission and an apparently perpetual lightning rod for any form of opposition, Clinton received all the blame for their defeat from the supporters of Black Rock. The *Black Rock Beacon*'s diatribe in March 1823 was typical: "We hope the senate [will] destroy, root and branch, everything that savors of Clintonianism."[7] It was an effort to overcome this kind of black environment that had motivated Clinton and his friends to generate such fanfare at the opening of the canal into the Hudson River in October 1823.

To little avail. Clinton may have been dreaming happy dreams about John Quincy Adams, Thomas Jefferson, and juicy rumors promoted by young Albany lawyers, but the presidential campaign was going to affect him in ways that would turn his dreams into a nightmare. As nothing touched him without touching the Erie Canal at the same time, the unfolding political drama requires our attention.

* * *

In 1824, the Electoral College that officially elects the president of the United States had not yet become the archaic symbol it is today. Discretion in voting was not unusual and response to outside influences was to be expected. Although by that time the electors were chosen in eighteen states through direct election by eligible citizens, New York was among the six states where they were still elected by the state legislature, with the members meeting "in caucus" to choose their candidate; De Witt Clinton was one of the most outspoken opponents of this clearly undemocratic system. Martin Van Buren had been representing New York in the U.S. Senate since 1821, but the state legislature remained under the control of his Bucktails and his hand-picked leadership known as the Albany Regency—a group characterized as "formidable in solidarity."[8] The events that followed would repeatedly demonstrate the accuracy of this portrayal.

On St. Valentine's Day 1823, the Van Burenite majorities in the New York legislative caucus nominated William Crawford as the Republican Party's candidate for president, as well as Albert Gallatin for vice president. Crawford was a respectable choice. A fifty-two-year-old Virginian, he had over twenty years' experience in politics. He had been a U.S. senator, minister to France, and James Madison's secretary of war. In 1816, he had been in the running for the presidency itself but stepped aside in favor of James Monroe. In return, Monroe appointed Crawford secretary of the treasury, a position he filled with distinction. Crawford's primary attraction for the Bucktails was his support for the contentious legislative caucuses rather than the popular elections endorsed by Clinton and his contingents.

The Clintonian, anti-caucus, and anti-Crawford forces were divided among the supporters of John Quincy Adams, General Andrew Jackson, and Speaker of the House Henry Clay of Kentucky (who was among the greatest enthusiasts for public improvements like the Erie Canal). But Adams, Jackson, and Clay were in agreement that the caucus system and Van Buren's power play were undemocratic. In April 1823, they merged their supporters into the People's Party, which included both supporters of Clinton and anti-Clintonians. Clinton leaned toward Adams, but not because of any sense of personal affection—Adams matched Clinton's arrogance and cold manner with associates as well as with opponents. Rather, Clinton admired Adams as an anti-slavery man of the north who supported internal improvements and cultural institutions.

Early in 1824, Governor Joseph Yates called for a special session of the New York state legislature to take place in August to change the electoral system, asserting that the people were concerned about "*their undoubted right*" of choosing electors.[9] Yates showed rare courage in choosing to run directly into the fierce opposition of the Bucktails. Clinton, who supported the decision, nevertheless wrote to a friend that "the executive is hors de combat—Wotan's horse, Balaam's Ass, Livy's ox and Mahomet's Camel all harnessed together could not draw him out of the kennel of public indignation."[10] There were even supporters of the caucus system who claimed Yates had called for the special session simply to play to the electorate, with full knowledge that any positive action would come too late to matter in the forthcoming elections.

Through it all, the Albany Regency's dominance held firm. When the time came, Yates's special session would produce nothing but bitterness and pledges of revenge from the losing side. Indeed, the Regency later succeeded in ousting Yates as candidate for governor in the upcoming election, substituting Samuel Young in his place. Young was a more outspoken anti-Clintonian, even though he had been a member of the canal commission since 1816. The bitter sentiments were reciprocated: as we have noted, Clinton had described Young on one occasion as "a blackguard [guilty of] audacious calumnies."

The People's Party chose James Tallmadge Jr. as its national leader. Tallmadge was immensely popular. He had been a general in the War of 1812, a congressman in 1817–1819, and a firm supporter of the Erie Canal, and was now, not incidentally, a member of the state Assembly, with his own ambitions to be governor of New York. While serving in the House of Representatives in 1819, Tallmadge had introduced a bill to deny Missouri statehood unless "the further introduction of slavery be prohibited in said State of Missouri, and that all children born in the State after its admission to the Union shall be free at the age of twenty-five years." Although this proposal was ultimately defeated, it led to the Missouri Compromise of 1820.

* * *

While all this was going on, the Albany Regency perceived the controversy surrounding the electoral process as a perfect opportunity to simultaneously accomplish three objectives that would make their own dreams come true: rid themselves of De Witt Clinton and his presidential pretensions once and for all, split the People's Party and thereby ruin Tallmadge's plans to be elected governor, and divide the anti-Crawford forces. Success would assure New York's forty-six electoral votes for Crawford, providing a high probability of his election as the next president of the United States—and beholden to Van Buren and his followers for their achievements in his behalf.[11]

The scheme was simplicity itself. The Bucktails planned to introduce a resolution in the final moments of the regular legislative session in April 1824 to strip Clinton of his last remaining link to political leadership—his membership on the canal commission, of which he was still the president. Any members of the legislature tempted to vote against the Regency by opposing Clinton's removal would be labeled as embracing a has-been nobody wanted.

It was a sign of the times that a step as audacious as this could succeed. Clinton had been on the canal commission from its inception in 1810 and president since 1816. During all those years, he had devoted a large proportion of his time to the canal without monetary compensation of any kind. Although he was given to denigrating the important contributions of others to the canal's success, he was the one who stood up most visibly, and most bravely, with the most to lose from the storms of opposition and incredulity. Every day, progress in canal construction and the rising flow of toll revenues testified to his stubborn optimism and vision. Toll revenue in 1824 was already sufficient to cover half the debt service—and the stretch from Rochester to Lake Erie was not yet open. In addition, Clinton was from many standpoints the best-known New Yorker on the national scene and already legendary for his accomplishments. Van Buren had noted in his autobiography that Clinton "did not lack troops of devoted personal adherents," but that type of devotion was insufficient.[12] As Van Buren observed, Clinton had failed to generate widespread popularity despite the pervasive enthusiasm for his crowning achievement, the Erie Canal.

After taking everything into consideration, De Witt Clinton's enemies were now prepared to take the risk of heaving him into political oblivion. With Van Buren in Washington, the Bucktail leader in the state Senate was Judge Roger Skinner, one of Clinton's longtime and outspoken adversaries. The procedure was carefully planned and executed. Its consequences were not.

On April 12, minutes before the legislature was about to adjourn until November, State Senator John Bowman of Monroe County rose to introduce a resolution to remove Clinton from the commission. Without debate, the resolution went immediately to a vote and passed 24–3. The results were met with total silence for what seemed like an endless period of time. Then the Senate adjourned for the season and filed out of the chamber.

The Assembly was almost but not quite as passive about the deed the Senate committed. After the resolution was read, but before the vote, Henry Cunningham of Montgomery County rose to speak in Clinton's support. Cunningham had been taken totally by surprise—he was in the process of putting on his overcoat when the vote results were announced—and had to struggle to gain attention for his remarks. Although Cunningham was not as well educated as some members of the Assembly, he was highly intelligent with a natural charm that made him an effective legislator.

Cunningham's peroration was an elaborate and comprehensive denunciation of the destructive deed carried out by Clinton's enemies, especially impressive for having been delivered ad lib. Cunningham began by emphasizing his political independence: "However much I esteem Mr. Clinton as a profound statesman and scholar, I am not embarked in his political fortunes, but free and untrammeled without fear, favor, or affection." After a lengthy recitation of Clinton's achievements, Cunningham was at his best when he placed the burden of the consequences on the shoulders of those responsible for the action: "I hope there is yet a redeeming spirit in this house—that we will not be guilty of so great an outrage. . . . What, let me ask, shall we answer in excuse for ourselves, when we return to an inquisitive and watchful people? . . . What can we say [Clinton] has been guilty of, that he should be singled out as an object of state vengeance? Will some friend of this resolution be kind enough to inform me? . . . This resolution was engendered in the most unhallowed feelings of malice, to effect some nefarious secret purpose at the expense of the honor and integrity of this legislature." Cunningham then forecast with remarkable accuracy, "When . . . the political bargainers and jugglers, who now hang round this capitol for subsistence, shall be overwhelmed and forgotten in their own insignificance . . . the pen of the future historian, in better days and in better times, will do [Clinton] justice, and erect to his memory a proud monument of fame as imperishable as the splendid works which owe their origins to his genius and perseverance."[13]

Two other assemblymen rose to speak briefly in Clinton's support and then the vote was taken. The result was 64–34 in favor of the resolution removing Clinton from the canal commission. James Tallmadge voted with the majority against Clinton, a step he would soon regret.

After fourteen years Clinton was no longer a canal commissioner. Other commissioners served longer terms—for example Stephen Van Rensselaer, a member of the original group, served until 1839. Clinton's enemy Samuel Young was a commissioner from 1816 to 1835, and William Bouck, an obscure politician until he became governor in 1843, served from 1821 to 1840.

* * *

This extraordinary event came as an electric shock to the entire state. Van Buren saw it the same way. The vote, he recalled in his autobiography, gave Clinton what up to that time he had so sorely lacked: "the sympathies of the People."[14] Indeed, popular excitement was immediate and powerful. Mass meetings were held in Albany and in New York City. The meeting in New York was so large that no public hall was spacious enough to hold the crowds. The ten thousand protesters finally gathered for their "indignation meeting" in City Hall Park, not far from the headquarters of Tammany Hall; one participant thirty years later recalled the assembly as "tens of thousands."[15] Twenty-five distinguished citizens of the city were dispatched to Clinton's Albany residence to report to him on the fury over the legislature's coup. Many smaller communities were equally aroused. Upstate in Canandaigua, citizens even considered organizing a physical attack on the legislators who had voted in favor of removal.[16]

Thurlow Weed, an articulate journalist from Rochester and one of Clinton's most devoted admirers, claimed that "the feeble and impotent attempts of blind and stupid malevolence . . . serve only to confirm and strengthen his claims."[17] A typical response from the public appeared in the *Albany Gazette* of April 20 in a communication signed by FIAT JUSTICIA; it presented an extraordinarily elaborate Corinthian column, "erected in the honor of one and the infamy of many," 8½ inches long and a full column wide as printed. The column showed "D.W.C." on its base and was followed by an announcement that "This MONUMENT is erected in testimony of a NATION'S GRATITUDE for the eminent services of that distinguished benefactor of mankind DE WITT CLINTON . . . who has added imperishable lustre on his own age, and conferred innumerable blessings on POSTERITY."

Even with all this popular clamor, Clinton was not a certain choice to run for governor on the ticket of the People's Party. He remained a controversial figure despite the crude and inappropriate attack by his enemies. But after some vacillation back and forth, the leaders of the party finally decided that the popular pressure was too great to deny him the nomination for governor. Yet the result fell well short of acclamation: the nominating convention gave seventy-six votes to Clinton, to Tallmadge's thirty-one—he ended up as candidate for lieutenant governor. Tallmadge had been so bold as to vote with the majority on the resolution for Clinton's removal. Now he and Clinton would be frosty but inseparable companions for the next few months of campaigning. The People's Party also came out for John Quincy Adams for president, to the satisfaction of Clinton (who had also been lending some support to Jackson) and the despair of former Federalists who considered Adams a liberal traitor to his father's conservative causes.

The campaign between De Witt Clinton and Samuel Young for governor was a lively one.* As early as April 15, the *New York Post* offered this description of what went on: "The envenomed malignity . . . must cause the cheek of every honorable man who calls himself a New Yorker to glow with a blush of shame and indignation." A colorful—but by no means exceptional—example of envenomed malignity appeared in the *Albany Daily Advertiser* on October 19: "[The Regency] will weep and wail and gnash their teeth, when the storm of popular fury shall overtake and utterly destroy them."[18] On the other side, Tammany, rallying its squaws and papooses, accused Clinton of sins and treason against his own party.† To this they added their usual refrain, "He is haughty in his manners and a friend of the aristocracy—cold and distant to all who cannot boast of wealth and family distinctions."[19] Nevertheless, Clinton and Young made some effort to preserve the niceties; at the end of a particularly vociferous letter to Clinton, Young was careful to sign off "with due respect."[20]

This campaign was especially interesting because the New York constitutional convention of 1821 had removed the freehold qualification from the franchise and large numbers of men were now in a position to vote for the first time. The process of nominating candidates by party conventions was also established, with Utica and Herkimer serving for many years as the meeting places of the parties.

The state election of November 1824 generated much excitement. Added to the change in suffrage, this resulted in a turnout of 190,545 voters, the highest for New York State up to that time, double the number who voted in the 1820 election, and 73 percent of the eligible electorate. When all the votes were counted, Clinton had defeated Young by a majority of 16,359 with 103,452 votes. Clinton's margin four years ear-

*Except for the tragic death of Clinton's sailor son James on July 14, at sea, from yellow fever.

†Meanwhile, according to the Tammany historian Gustavus Myers, under Tammany management, "the streets [of New York City] were an abomination of filth. . . . As a result of the bad water . . . and the uncleanliness of the streets, yellow fever and cholera had several times devastated the city" (Myers, *The History of Tammany Hall*, p. 68).

lier had been only 1.6 percent of the total vote; in 1826, his margin of victory would be merely 1.9 percent of the total.[21]

More remarkable, and revealing, Tallmadge's majority over Erastus Root for lieutenant governor was 32,400. In Tompkins County, the future home of Cornell University at Ithaca, Tallmadge received 3633 votes, while Root's count was precisely 3. Clinton would take office with the support of three-quarters of the legislature.

The excitement was so intense that Van Buren's arrival at the polls on election day provoked angry shouts of "Regency! Regency!" and about a dozen people challenged the validity of the U.S. senator's registration as a voter. The election board and even some of Clinton's "most attached friends" failed to get the challenge withdrawn, and Van Buren had to take the prescribed oath, just like any first-time voter.

Although the result in the election for governor of New York was clear and beyond dispute, that day's presidential results were anything but. Crawford had been enthusiastically supported by the Bucktail representatives from New York, even after he suffered a serious stroke and was left paralyzed and nearly blind. The electoral vote was 99 for Jackson, 84 for Adams, 41 for Crawford, and 37 for Clay, leaving none of them with the necessary majority. Under these circumstances, the Twelfth Amendment provided that the House of Representatives should elect the president from among the three candidates receiving the most votes, which removed Clay from the contest. Clay, who was Speaker of the House, had sufficient influence to swing his supporters to vote for Adams, who then became the sixth president of the United States—and promptly appointed Clay secretary of state.

Adams also rewarded Clinton for his support (or tried to remove him from the scene) by offering him the ambassadorship to Great Britain. If Clinton had taken Adams up on his offer, being abroad would have provided time for the political clutter adhering to him to subside, greatly improving his chances to fulfill his dream of the presidency in 1828. But Clinton declined. The people of New York, he declared, had elected him with such a heavy majority that he was obliged to respond to their call. The response was gracious but perhaps less than the full truth of the matter. Power mattered. Appealing as the offer of ambassadorship may have been, in all likelihood a return to the governor's manion was the greater temptation.

Indeed, Clinton was now in his element. Shortly before his inauguration, he moved his family into a magnificent new home at the corner of North Pearl and Steuben streets in Manhattan. In a fine portrait at that moment by Samuel F. B. Morse, who would one day invent the telegraph and who was one of Clinton's close friends, we see a confident man with a large and well-formed head, brown hair carefully combed back, clear brown eyes, and a thin-lipped mouth grimly set and turning down at the corners.*

Clinton's inaugural address included an impressive recital of progress and prosperity on the Erie Canal—revenues exceeding loan interest by $400,000, functioning transportation all the way from the Hudson up to the final construction steps at Buffalo, and rapid growth in the demand for luxury items from the settlements along the canal. Quite aside from his focus on the canal, his administration included "An Act to Prevent the Inhuman Treatment of Slaves," aid to agriculture in the development of methods to improve productivity of both land and labor, steps for the betterment of conditions in the state's prisons, the establishment of normal schools to train teachers, and the construction of a state road to run parallel with the Erie Canal all the way from Albany to Buffalo.

* * *

It is interesting to speculate on Martin Van Buren's view of his own role in all this political turmoil. Like any politician whose future is in the hands of the electorate, Van Buren felt the pressure to sustain his popu-

*This portrait hangs in the American Wing of the Metropolitan Museum of Art in New York City.

larity. In September 1822, he had confessed to a friend, "Why the deuce is it that they have such an itching for abusing me. I try to be harmless, and positively good natured, & a most decided friend of peace."[22]

His autobiography does not stint in relating what happened in 1824 and his own reaction to it—although Van Buren did not begin writing this book until he was seventy-one, some thirty years after his party's demotion of Clinton. As a result, his account of the removal of Clinton as commissioner may be either inaccurate or colored by an effort to avoid tarnishing his own reputation. In any case, Van Buren deplores that "the motion came from our side—and ours was the responsibility." Indeed, Clinton "had confessedly done more than any other man to secure the success of the great Public Work [the Erie Canal]. . . . Having the best right to be regarded as the founder of the Work," his contribution fully entitled him to the mark of distinction as president of the canal commission.

Van Buren claims he "had no knowledge, being in Washington, of the intention to make [the removal]." He buttresses this claim to innocence by recounting that his "excellent friend" Judge Skinner had been in charge in Albany during his absence in Washington and was "more instrumental in accomplishing the removal of Mr. Clinton than any other of our friends." Furthermore, according to Van Buren, Skinner was well aware that "if informed of the design, I would have done what I could to prevent it." Accordingly, Skinner "took especial pains to keep it from me and laughed at the apprehensions I expressed on being informed of the act. He was standing at the window, tapping the glass with his fingers, whilst I was taking my breakfast with what appetite his news had left me. I could not resist saying to him—'I hope, Judge, you are now satisfied that there is such a thing in politics as *killing a man too dead!*'" Judge Skinner was wounded deeply by the violence of Van Buren's reaction and by "the pang inflicted on his heart by Mr. Clinton's success and by the conviction that he had contributed to it. . . . He died not long after in my arms."[23]

Van Buren's declaration of innocence and sorrow about his party's ruthless effort to eliminate Clinton recalls how Shakespeare describes a similar state of affairs in *Richard II*. Exton, one of the members of the royal court, announces to Henry Bolingbroke, the duke of Hereford, that he has done Bolingbroke the favor of murdering the duke's imprisoned cousin King Richard II, thereby making it possible for Henry to achieve his long desired—and clearly visible—ambition to ascend the throne as Henry IV. Here is how Henry responds to the news:

> They love not poison that do poison need,
> Nor do I thee: though I did wish him dead,
> I hate the murderer, love him murdered.
> The guilt of conscience take thou for thy labour,
> But neither my good word nor princely favor. . . .
> Come, mourn with me for that I do lament,
> And put on sullen black incontinent.

Despite his misgivings about killing a man too dead, Van Buren had not yet reached the pinnacle of his political career. After a brief stint as governor of New York in 1829, Van Buren resigned and returned to Washington to serve as secretary of state in Andrew Jackson's first term. He advanced to the vice presidency in Jackson's second term and then succeeded Jackson in the White House in the election of 1836, at the age of fifty-four (he would live to the age of eighty). But Van Buren's inauguration as president would be a bitter victory. He was no sooner in office than the nation was assailed by one of the worst financial panics and economic breakdowns in our history. He claimed the government was helpless before the crashing economy and financial system and called on the American people to take care of themselves rather than turning to government for assistance and bailouts:

> Those who look to the action of this government for specific aid to the citizen to relieve embarrassments arising from losses by revulsions in commerce and credit, lose sight of the ends for

> which it was created, and the powers with which it is clothed. It was established to give security to us all. . . . It was not intended to confer special favors on individuals. . . . The less government interferes with private pursuits, the better for the general prosperity.[24]

Almost a hundred years later, and in even more shocking conditions, Herbert Hoover would plant himself firmly in Van Buren's footsteps: "The evidence of our ability to solve great problems outside of government action . . . [and] victory . . . will be won by the resolution of our people to fight their own battles . . . by stimulating their ingenuity to solve their own problems, by taking new courage to be masters of their own destiny in the struggle of life."[25]

Like Herbert Hoover, Martin Van Buren was only a one-term president.

But in January 1825, while Van Buren was facing a future rendered uncertain by the Clinton debacle, Clinton was looking forward to his greatest moment of glory. In October, as governor of the State of New York, he would be the leading player in the high drama of his age: the completion and opening of the full Erie Canal. After all the grief, all the frustration, all the raw insults and denigrations, he was at long last at the peak of his public popularity.

CHAPTER 17

THE WEDDING OF THE WATERS

In response to an invitation from President James Monroe to be the first "Nation's Guest" of the United States, the Marquis de Lafayette landed in New York City in August 1824 for a triumphal American tour. And the tour was indeed a triumph.

Lafayette was the last surviving major general from the War of Independence and a widely popular symbol of opposition to monarchical tyranny. Although sixty-seven years old and in poor health, he traveled up, down, and across the United States for thirteen months, welcomed by countless thousands of shouting crowds, and transported by every available means of conveyance. He visited all twenty-four states of the union, surveyed both Harvard and Columbia universities, dined with all three living ex-presidents, the current president, and the next president, and sat through what must have amounted to hundreds of speeches, dinners, toasts, and other forms of celebration. At Boston, for example, he rode in a carriage drawn by six white horses as part of a giant parade of seven thousand citizens and led by survivors of the Battle of Bunker Hill, who presented him with some soil from Bunker Hill to take back to France.

On June 6, 1825, Lafayette arrived at Buffalo by steamer from

Dunkirk, fifty miles down the coast of Lake Erie. It was time to pay his respects to the Erie Canal. He went by carriage from Buffalo to Lockport, where he marveled at the great works nearing completion and at the aqueducts over which the canal "pursued an aerial route." And then he boarded the passenger boat that carried him on the canal from Lockport on to Rochester, Syracuse, and Utica. This boat was known as a packet, or packet boat, so named to evoke the same term applied to the great sailing ships carrying passengers back and forth across the Atlantic Ocean.

When Lafayette arrived at Utica on June 8, De Witt Clinton greeted him with a brand-new packet boat named *Governor Clinton*. Not to be outdone by Boston, Clinton had arranged for the boat to be pulled by white horses instead of the usual drab mules. Lafayette and Clinton spent the next three days sailing eastward on canal waters to Albany, where they passed under a triumphal arch topped with a large stuffed eagle that flapped its wings by some form of mechanism at the moment of their arrival.

None of these ornate ceremonies commemorating the Revolution and America's independence could rival the celebrations, revels, festivities, boasts, and toasts that marked the official completion of the Erie Canal just four months later, in October 1825. All the frictions, rivalries, and bitterness of the years past would be forgotten in the avalanche of lavish entertainments and ceremonies that consumed New York State as it greeted a dream coming true.* New York was entitled to go all out. In the space of just eight years, with no financial or any other kind of assistance from the national government or any sister state, New York had created what the official chronicler of these events, William L. Stone,

*There are many descriptions of this extraordinary occasion. With minor exceptions such as contemporary newspaper commentary, all sources have drawn upon a remarkable eyewitness "narrative" of the opening of the canal by William L. Stone, which was prepared at the request of the Committee of the Corporation of the City of New York. All quotations in this chapter are from Stone—accessible at www.history.rochester.edu/canal/bib/colden/App18.html—unless otherwise indicated.

aptly described as "a work of which the oldest and richest nations of Christendom might well be proud."

The celebration began on the morning of Wednesday, October 26, and continued on almost uninterruptedly into the night of Friday, November 4, followed by a huge ball on the evening of November 7. The most lavish impresarios of our own time, from Billy Rose to Franco Zeffirelli, never staged anything as elaborate or as prolonged as what came to be known as the Wedding of the Waters.

The most impressive feature of the whole series of events was the stamina of the participants, especially De Witt Clinton, now in his role of governor, and Lieutenant Governor Tallmadge, who were front row center from the very beginning through stops for ceremonies at more than twenty towns right up to the final conclusion of the celebrations. The long nights trying to sleep in the cramped quarters of their boat could hardly have been refreshing. One can only be in awe of the number of speeches Clinton and Tallmadge had to listen to, the number of speeches they had to deliver, the number of toasts they had to drink, the number of feasts they had to consume, and the number of blasts of artillery that assailed their ears at nearly every stop and nearly every departure.

Buffalo opened the festivities at nine o'clock on the morning of October 26 with a parade led by Clinton from the courthouse to the waterfront. There Clinton and Tallmadge boarded their home for the next eight days, the packet *Seneca Chief*, drawn by four gaily decorated gray horses and carrying two elegant wooden kegs decorated with eagles and filled with water from Lake Erie to be poured into the Atlantic Ocean upon arrival in New York City. The cabin of the *Seneca Chief* contained two paintings, one a view of Buffalo and its junction with the canal, but the other far more elaborate. Although the artist was a well-known miniature portrait painter, his work here was massive. It showed Hercules resting from his labors while Clinton, full length and in Roman costume, is inviting Neptune to join him. Neptune is erect in his seashell, drawn by sea horses and surrounded by Naiades—but both Neptune and his companions appear astonished at the canal lock opening ahead of them.

The *Seneca Chief* was followed by the *Superior*, the *Commodore Perry*, a freight boat, and the *Buffalo*. A long line of other boats came along behind these first five. Among these was the *Noah's Ark*, bringing east a motley cargo of fauna from the west, consisting of birds, beasts, "creeping things," a bear, two eagles, two fawns, several fish, and two Indian boys in the dress of their nation. The boat bringing up the end of the parade—as far away from Clinton as possible—was the *Niagara* of Black Rock, with Peter Porter aboard.

The first event before the boats were boarded was a brief address by Jesse Hawley, the man who had started the agitation for a canal across New York State twenty years earlier. It was Hawley who wrote back then, "If the project be but a feasible one, no situation on the globe offers such extensive and numerous advantages to inland navigation by a canal, as this!" Hawley now set the tone for the occasion by paying his respects to the "projectors . . . statesmen . . . legislators . . . engineers [and] men who had executed this magnificent work—an exhibition of the moral force of a free and enlightened people to the world." But he could not resist a dig at those who had ignored his predictions for so long: "This is all the notice I have ever received from the State of the people of New York for it in any wise."[1] Yet three years earlier, De Witt Clinton himself had written to Hawley admitting that the first suggestion he had seen of a canal from Lake Erie to the Hudson had been in Hawley's essays.[2]

As the *Seneca Chief* pulled away from the docks, there were simultaneously a rattle of small arms fire, a blast from a thirty-two-pound cannon, the band playing *fortissimo*, and robust cheers from the crowds on the shores answered by roars from the men packed aboard the line of boats about to sail off on the Erie Canal to New York harbor. That was just the opening sally. New York and the communities in between were impatiently awaiting word that the *Seneca Chief* was under way.

The news was signaled by artillery lined up all the way from Buffalo to New York, with each gun within audible range of the other guns on either side of it. As each shot was fired, the next gun went off, all the way from Buffalo to Sandy Hook, New Jersey, at the southern end of New York harbor, at which point the guns fired a return sequence back to Buffalo. Many of these guns had been British arms captured by Oliver Hazard Perry at the Battle of Lake Erie, and one of the gunners had served as a lieutenant in Napoleon's army. The round trip of cannonades took more than two hours (Clinton had hoped the procedure would provide a measurement of the speed of sound, but the planned arrangements failed to produce the answer). As the last returning shot boomed, the line of boats departed for Lockport—while the Buffalonians carried on their revelries well into the night.

The guns set off emotions as powerful as their explosions. Cadwallader D. Colden recalled the very moment he heard the cannon and asked himself, "Who that has American blood in his veins can hear this sound without emotion? Who that has the privilege to do it, can refrain from exclaiming, I too, am an American citizen; and feel as much pride in being able to make the declaration, as ever an inhabitant of the eternal city felt, in proclaiming that he was a Roman."[3]

* * *

The parade reached Rochester on the rainy morning of the twenty-seventh. In addition to the usual bombardments and speeches, Rochester provided a new boat to join the parade, carrying Rochester's favorite description of itself, the *Young Lion of the West*. This boat waited as the others approached and then hailed the lead boat with the cry "Who comes there?" and the response came, "Your brothers from the West, on the waters of the great Lakes." The dialogue continued until the *Young Lion of the West* gave way and allowed the brethren from the west to enter Rochester's spacious harbor, where the side of the canal was lined with every possible facility for boats to dock and load or unload their cargo and passengers. Few boats have ever enjoyed such a special kind of reception: the *Young Lion of the West* was carrying two wolves, a fawn, a fox, four raccoons, and two eagles, all creatures of the Rochester area.

The smallest villages the flotilla passed through were determined to create celebrations as elaborate as anything the larger towns could put on. Take, for example, Port Byron, a settlement not far from Syracuse and then known as Bucksville after its founder, Aholiab Buck. Even today, Port Byron has a population of only 1400 and an official area of merely 0.9 square miles.* Yet this tiny community greeted Clinton and his companions with a deafening fireworks display and musket volleys, followed by the launching of an illuminated balloon, which obligingly took an easterly direction and floated along the line of the canal. Then came a full-fledged banquet, including a fat ox roasted whole, and a succession of toasts whose length was not to be exceeded in most larger towns along the route. All that was only a prelude, as "thirty or forty ladies, always patriotic, [and] arrayed in their sweetest smiles and most beautiful attire, awaited the happy moment when they could 'trip the light fantastic toe' with the expected strangers."

On Saturday, at Syracuse, another early supporter of the canal led the reception—Joshua Forman, who in February 1808 recommended to the legislature the establishment of a joint committee to arrange for exploring and surveying the most favorable route for a canal between the Hudson River and Lake Erie. That proposal had been just the first step in the tortuous process of nine years before the state legislators would finally authorize the project they were now outdoing one another in proclaiming the masterwork of the age.

* * *

*Port Byron has had its fair share of famous residents: Henry Wells, the founder of American Express, the great Mormon leader Brigham Young, and Isaac Singer, who invented the sewing machine.

There were two notable killjoys among the celebrations. At Rome and Schenectady, the citizens claimed their towns had been short-changed by the design of the canal and expressed their displeasure by a notable absence of toasts, gunfire, cheers, and feasts.

At Rome, the Western Inland Lock Navigation Company had built a canal in the 1790s right through the middle of town, but the engineers on the Erie Canal determined that a route running along the outskirts would be more efficient. The frustrated Romans were convinced the canal commissioners had treated them unjustly, and they took the occasion of the Wedding of the Waters to make their sentiments clear. At eleven o'clock on the morning of Saturday, October 29, they formed the usual uniformed parade in front of the usual hotel with the usual bands and the usual artillery activity, but the rear of the parade featured four men supporting a black barrel filled with water from the old canal. Marching to muffled drums, the four men escorted their barrel to the new canal, where they unceremoniously dumped its contents.

Having expressed their view of the matter in such eloquent fashion, the citizens of Rome turned around and joined in the spirit of the lively celebration under way. A large throng gathered at the hotel for dinner and toasts, including a toast for the Erie Canal, "to whom honor is due." When the flotilla from the west arrived on Sunday, Rome's city fathers received De Witt Clinton and the other officials with "the usual courtesies"—but the visit lasted only about an hour, then the boats departed to a more gracious reception at Utica.

Schenectady had similar complaints about the canal's route. The leading newspaper even proposed greeting the Clinton party with a funeral procession or some other demonstration of mourning. Nobody went that far, but Schenectady made no preparations for a large-scale reception. When the boats came into sight about three o'clock on the afternoon of Tuesday, November 1, the principal citizens respectfully welcomed the governor and lieutenant governor and provided them with a dinner at the main hotel, but the crowds were silent. The occasion developed into a kind of replay of "the gloomy interval" the commissioners had experienced in Schenectady in 1810. An hour later, the company reboarded the boats and, in a dark and dreary night, continued on their way into the formidable sequence of locks by which the canal descends into the Hudson valley.

* * *

At 10:30 a.m. the next morning, Wednesday, November 2, Clinton and his companions arrived at Albany and passed through the last lock of the Erie Canal before entering the Hudson River. Just a week had elapsed since the explosive departure from Buffalo—and the guns here were just as busy: twenty-four cannon on the pier fired a grand salute as the *Seneca Chief* and her companions left from the last lock and entered the basin leading to the Hudson. The *Albany Daily Advertiser* caught the spirit of Clinton's arrival in Albany: "It was not a monarch which they hailed, but it was the majesty of genius supported by a free people that rode in triumph and commanded the admiration of men stout of heart and firm of purpose."[4]

The Albany basin was jammed with canal boats and a huge gathering of cheering spectators massed along the wharves, the bridges, and the shoreline. After the line of boats reached the southernmost bridge across the basin, the Clinton contingent went ashore to be received by a welcoming committee that included every available local official and even delegates from the national government in Washington, including Secretary of State Henry Clay, Chief Justice John Marshall of the U.S. Supreme Court, Attorney General William Wirt, and high-ranking military men.

Accompanied by a long parade featuring carts loaded with western produce, the honored guests and their innumerable hosts walked through Albany to the Assembly chamber. Here the scene was highlighted with portraits of both De Witt Clinton and his uncle George, as well as an enormous full-length portrait, *The Father of his Country*, of

George Washington, on top of which a carved bird of victory grasped a shot of lightning. A large chorus and full orchestra proceeded to perform as a prelude to the inevitable speeches. The keynote, by a man named William James, contrasted the primitive conditions in New York before 1817 with the striking improvements over the past eight years. James's vivid description was full of flowery language, like "the dismal and savage trackways . . . through forbidding forests, where now stand . . . flourishing towns . . . celebrated for the elegance and refinement of their inhabitants, the grandeur of their scenery [and] seats of learning."

Then everyone returned to the bridge, which had been converted into a gothic cathedral, with pointed arches 14 feet high and pilasters capped by gilded gothic turrets. Elaborate lines of shrubbery decorated the sides of the bridge and the arches. Farther on, there were three circular arches topped by huge signs reading "GRAND ERIE CANAL," "JULY 4TH, 1817," and "OCTOBER 26TH, 1825." A vast tent stretched beyond these arches, containing two lines of tables, each 150 feet long, to accommodate six hundred guests.

The aisle was wide enough for another procession to march through the area. And then the guests could finally be seated and begin to enjoy "plenty of the 'ruby bright' wines of the best vineyards of Europe," which must have been mighty welcome by that time. The festivities continued on into the evening, when there was an elaborate theater performance of odes, a full drama, and a canal scene with locks, including horses and boats actually passing across the stage.

* * *

Thursday morning turned out to be one of those gleaming and luminous days, ideal for blessing the very first boats about to complete the voyage by uninterrupted waterway from Buffalo to New York City and the Atlantic Ocean beyond. The crowds along the shores at Albany were even larger than on the preceding day, with every dock, store, and vessel full of shouting multitudes.

The nautical procession to New York got under way at ten o'clock. Here there were no towpaths, so the canal boats were pulled down the river by a line of seven steamboats decorated with colorful banners and streamers fluttering in the wind, each carrying a brass band and overflowing with passengers. William Stone, a man of his own time rather than ours, greatly admired "the large columns of steam rushing from the fleet, rising majestically upwards, and curling and rolling into a thousand fantastic and beautiful forms."

While the steamer *Chancellor Livingston* took the *Seneca Chief* in tow, the *Saratoga*, a small steamboat capable of higher speeds than the others, served as a tender to transport passengers of the other boats to and from the landing places. Between stops, the *Saratoga* "sported about like a dolphin—now in the wake of one boat, now along the side of another, and now shooting a-head of the whole." As the flotilla moved down the Hudson, additional small steam vessels joined the *Saratoga* in darting back and forth and around the majestic line of boats heading to the city, until a total of twenty-two gaily decorated steamers had joined the procession.

As the naval parade sailed down the broad river, with its high banks covered by the rich foliage of autumn, they were greeted by the usual muskets, the cannon carefully placed to signal their arrival, the bands, and the shouts. Evening was approaching as they passed Hyde Park, where bonfires, rockets, and other illuminations spiced the standard trappings of the reception. At West Point, the greeting was even noisier, with a blast of rockets plus a salute of twenty-four guns upon the arrival of the first boats and another volley of twenty-four guns when the last boat passed.

Once West Point was behind them, and all the official greeters had departed, the passengers aboard the boats were finally able to retire and get some rest. They would need it: the following day, Friday, November 4, they were scheduled to arrive in New York City waters before dawn to commence the climactic festivities of their voyage from Buffalo.

When the flotilla reached Ossining, about thirty miles from the northern end of Manhattan Island, they were met by a brand-new steamboat, the *Washington*, chartered for the occasion. The entire stern area of the ship was covered with the most elaborate kinds of decorations, flaming torches, and sculptured figures celebrating George Washington, the Marquis de Lafayette, agriculture, commerce, and even the whole globe of the earth.

A group of New York City officials stood in the bow, and one cried out to the *Seneca Chief*, "Whence come you, and where are you bound?" The answer came back: "From Lake Erie—and bound for Sandy Hook." Now the boats started ahead and immediately encountered two British warships flying the American flag along with the usual Cross of St. George. An exchange of salutes by gunfire acknowledged the occasion.

By 8:30 a.m., the distinguished guests on the canal boats and their escorts joined the city officials at City Hall and proceeded immediately from there to the line of steamships at the foot of Whitehall Street, where the trip into the harbor would begin. One of the steamers involved was the *Lady Clinton*, an "elegant safety-barge" decorated with so much foliage it must have looked like a small forest.* The *Lady Clinton* was reserved for the ladies of the party. According to Stone, the captain "paid every attention to his beautiful charge; every countenance beamed with satisfaction, and every eye sparkled with delight."

After stops at the navy yard on the East River and greetings from the crowds jamming the shores of Brooklyn Heights, the steamboats sailed past Castle Garden at the Battery and headed out toward Governors Island and Jersey City, and then, in calm seas and brilliant sunshine, sailed on out to Sandy Hook, at the very southern end of New York harbor. Here Clinton, Tallmadge, and a long procession of other dignitaries transferred to the *Washington*, while all the other boats and ships of varying sizes and shapes formed a great circle around her.

*This was the same packet boat Clinton had seen passing along in May 1822, as noted with special pleasure in his diary at that time.

As the very first act of the ceremony, Clinton filled several bottles—noted as "made in America"—with the water from Lake Erie. Then he placed them in a cedar box specially prepared for the occasion by the famous woodworker Duncan Phyfe, to be transported back to France as a gift to the Marquis de Lafayette from the people of New York.

Clinton then performed the culmination, not just of the ceremonies on that day, but of all the years of hope and anger, progress and retreat, and design and redesign that led up to this moment: from the green keg with gilded hoops, he poured the Erie waters into the Atlantic Ocean. Using many fewer words than usual, he turned toward his companions and declared:

> The solemnity, at this place, on the first arrival of vessels from Lake Erie, is intended to indicate and commemorate the navigable communication, which has been accomplished between our Mediterranean Seas and the Atlantic Ocean, in about eight years, to the extent of more than four hundred and twenty-five miles, by the wisdom, public spirit, and energy of the people of the state of New York; and may the God of Heavens and the Earth smile most propitiously on the work, and render it subservient to the best interests of the human race.

At that point, Dr. Samuel Mitchill, one of Clinton's close friends, stepped up with thirteen bottles of water—one each from the Ganges, the Indus, the Nile, the Gambia, the Thames, the Seine, the Rhine, the Danube, the Mississippi, the Columbia, the Orinoco, the Rio de la Plata, and the Amazon—which he proceeded to empty into the Atlantic. Dr. Mitchill was followed by Cadwallader Colden, the extraordinarily verbose but distinguished grandson of his namesake, the man who had visited the Mohawk valley and farther west in 1724 and had enthused over the potentials there for east-west passages by waterway. Colden was there to give the mayor of New York his compendium on the history of canals and waterways, with special emphasis on the Erie Canal, a long but valuable history.

There were some present at these ceremonies who had attended a naval fête given in 1815 by the Prince of Wales on the Thames for the sovereigns of Europe in celebration of the defeat of Napoleon. According to these men, as Stone describes it, the spectacle in the waters of New York "so far transcended that in the metropolis of England as scarcely to admit of a comparison."

While all this was going on out in the furthest reaches of the harbor, a giant parade was under way in Manhattan. As hawkers came out in force selling Clinton "kerchiefs" and Clinton hats, the procession began on Greenwich Street, moved through Canal Street, proceeded up Broadway to Broome Street and up to the Bowery at its farthest point, after which it turned back down and ended at City Hall. At its maximum, the line of participants was more than a mile and a half in length, the largest parade ever witnessed in America up to that time (and maybe since). In addition to military contingents, just about every social, occupational, and religious group in the city participated—bakers, tailors, sailors, teachers, and even a large representation from Clinton's alma mater, Columbia University—each aiming to outdo the others in the elaborate character of their badges and in the beauty of their banners. Some groups had gigantic floats covered with rich Turkish or oriental carpets, with members of the sponsoring organization displaying how they pursued their daily activities.

As darkness fell, all the public buildings and main hotels were covered with brilliant illuminations, of which the brightest seems to have been the City Hotel on Broadway. That was most appropriate. The City Hotel had been the site of the mass meeting in December 1815, when the citizens rose up and instructed Clinton to prepare the memorial, insisting that the legislature authorize the construction of the Erie Canal to begin. There might never have been a canal without that meeting.

* * *

The citizens of Buffalo were not to be outdone by their fellow Americans at the eastern end of the canal. They had planned to follow the

New York ceremony by performing it in reverse. Early on the morning of November 23, the *Seneca Chief* completed a triumphant return voyage on the canal, bearing a keg of Atlantic waters westward to Buffalo. At 10:00 a.m., Judge Samuel Wilkeson stepped to the bow of the boat to do the deed. Wilkeson had earned the honor as the most aggressive of Buffalo's aggressive citizens in the struggle to be named the terminus of the canal. He emptied the keg of Atlantic waters into Lake Erie amid the cheers of the crowd and the explosions of the guns.

The Wedding of the Waters was now complete. Stone's final paragraph to his narrative of these proceedings was right to the point:

> The authors and the builders—the heads who planned, and the hands who executed this stupendous work, deserve a perennial monument; and they will have it. To borrow an expression from the highest of all sources, "the works which they have done, these will bear witness of them." Europe already begins to admire, America can never forget to acknowledge, that THEY HAVE BUILT THE LONGEST CANAL IN THE WORLD IN THE LEAST TIME, WITH THE LEAST EXPERIENCE, FOR THE LEAST MONEY, AND TO THE GREATEST PUBLIC BENEFIT.

Weddings are only beginnings. America's "Mediterranean Seas" were now wedded to the Atlantic Ocean–a big step toward the northwest passage Henry Hudson had been seeking more than two hundred years earlier—but what was life like after the honeymoon and the glow of this extraordinary wedding had worn off? How well did the marriage settle down to the routines of daily activities? New York State had made an enormous investment in the Erie Canal in terms of labor and reputation as well as in money. The time had arrived to receive the payoff, to evaluate the returns, and to observe the changing future.

PART V

After the Wedding

CHAPTER 18

NO CHARGE FOR BIRTHS

The Erie Canal was an instant success. In 1826, the year after the Wedding of the Waters, about 7000 boats were operating on the canal, compared with 2000 boats when Lafayette toured New York in 1824 and 1825. Equally impressive, the canal commissioners could report that toll revenue was running in excess of $500,000, five times the interest due on the canal's outstanding bonds.[1] That was just the beginning. In 1837, the commissioners reported that the entire debt was repaid. This was one of the commissioners' happier tasks as they continued to administer the canal in the years ahead.

But the swelling flow of freight and passengers was just part of the story. The canal became one of the wonders of the world. Travelers from all over the United States and Europe arrived to taste the enthralling experience of water travel, uphill and downhill, over nearly four hundred miles through dark forests and burgeoning towns. Expressed in the concepts of today's world, people perceived the canal as a combination of Disneyland, the Grand Canyon, and a high-tech laboratory in Silicon Valley. Literary and theatrical celebrities like William Cullen Bryant, Edward Everett Hale, Nathaniel Hawthorne, and Fanny Kemble all came to see the miracle of the age and pass judgment on it.

There is something wonderfully seductive—almost magical—about being pulled along an artificial river by a team of mules or horses.[2] The gliding ride has a quality of calmness and quiet that no motorboat or sailboat can match. Nathaniel Hawthorne, touring the canal in 1835, was inspired to describe the boat ride in these lofty words: "Behold us, then, fairly afloat, with three horses harnessed to our vessel, like the steeds of Neptune to a huge scallop-shell, in mythological pictures. Bound to a distant port, we had neither chart nor compass, nor cared about the wind, nor felt the heaving of a billow, nor dreaded shipwreck, however fierce the tempest, in our adventurous navigation."[3]

Hawthorne goes on to make a remarkable prediction: "Surely, the water of this canal must be the most fertilizing of all fluids; for it causes towns—with their masses of brick and stones . . . to spring up, till, in time, the wondrous stream may flow between two continuous lines of buildings, through one thronged street, from Buffalo to Albany." Although most of the celebrities visiting the Erie Canal, and especially those from Britain, tended to be patronizing about the little towns grouping along the route of the canal, these towns soon had educational facilities and theaters as advanced as any in the country. Indeed, the canal itself set off a long sequence of plays, poems, and novels, serious as well as comic.*

Hawthorne also refers to "all the dismal swamps and unimpressive scenery, that could be found between the great lakes and the sea-coast." That description fell far short of a full description of what travelers could admire as they glided through long passages between steep, rocky hills and the wild country around Little Falls as well as many charming valleys with well-tended farmlands. One more generous visitor portrayed the scenery as "the perfect Eden of a poet's heaven."[4] And then there was the marvel of the locks, bypassing rapids and waterfalls while scaling the heights and gently descending the downslopes; in addition,

*An outstanding book on this entire subject is Lionel Wyld's *Low Bridge!*

as another traveler pointed out, "the passenger can supply himself with provisions and grog at all the lockhouses along the line at a very low rate."[5]

John Howison, the English tourist who had visited the area in 1820, had been "particularly struck with the elegance and magnitude of the villages, and [one] often feels inclined to ask where the labouring classes reside; as not a vestige of the meanness and penury . . . is to be discovered."[6] But that pristine landscape was changing. In response to the mushrooming growth, once sleepy communities along the canal were coming to life with smoky factories and clattering mills, polluted air, slumlike living quarters, and other early symptoms of the Industrial Revolution. Howison himself had complained about the "rattling of mills, and the smoke of iron founderies" in Rochester.

When Howison was invited to wash up in the bar of a tavern beside the canal, he decided "I would rather have worn the coat of dust I received in the stage, than attempted ablution in [that bar]."[7] And a tourist approaching Buffalo in 1829 grumbled that the drinking water "which I tasted, I never wish to taste more, as it set my bowels in an uproar prodigiously, to my great inconvenience and pain."[8]

Despite the discomforts, and far more important, the canal delivered on its promise of speed. Its impact on travel and the movement of goods was spectacular. In 1810, De Witt Clinton and his fellow commissioners had needed thirty-two days to go from Albany to Buffalo. Now, in spite of the speed limit of four miles an hour—and even the most impatient drivers could not push their horses to pull at more than five miles an hour—passenger boats were making the trip from one end of the canal to the other in less than five days. Flatboats loaded with as much as fifty tons of freight took no more than six days. Traveling day and night, these boats slashed the costs of moving merchandise and commodities to only a small fraction of what wagon transport had charged, and in so doing vastly expanded the quantity and variety of merchandise that could profitably be brought to market.

* * *

Freight traffic on the canal was heavy from the start and generated the largest share of toll revenue. The typical boat, pulled by two horses or mules, weighed nearly fifty tons and was seventy-seven feet long. Its beam of fourteen feet three inches left just four and a half inches of clearance on each side as the boat passed through the locks. Despite Albany's strategic location at the eastern terminus of the canal, Albany was not a place to do much business. Instead, most of the freight boats, dozens at a time, were towed by steamers up and down the Hudson between New York City and Albany, and loaded or unloaded where the action was—at New York City. Robert Fulton had died in 1815, two years before the start of the canal, but in view of his great enthusiasm for canals, he would have been proud of the important role played by the steamboat in servicing the greatest canal created up to that time.[9]

The packet boats used by the more affluent travelers on the canal were more colorful than the boats transporting freight. The packets carried thirty to fifty passengers and no freight, ranged from sixty to seventy feet in length, and were pulled by three horses or mules. The passenger cabin occupied most of the space, but the packets also provided a kitchen and a bar, with small outside areas in the bow and stern for handling the boat and for the steersman. Line boats were less luxurious than packets and hauled some freight as well as passengers. Although many line boats were as long as eighty feet, they were pulled by only two animals instead of three, which made them slower than the packets. Their passengers, as a result, were not "so select a company."[10]

Although lengths and facilities varied, the fifteen-foot widths of the locks limited every boat on the Erie Canal to fourteen feet three inches in width—"four steps broad," as one traveler put it. The ceilings on the packet and line boats, however, were high enough so that gentlemen did not have to remove their ubiquitous top hats while inside; judging by the many illustrations of life aboard the packets, only bedtime would induce men to remove those hats—and maybe not even then.

The passenger cabin, on occasion with library facilities, was a busy place during the day, as passengers sprawled out to read, gathered to play games, or congregated in conversational groups. The claustrophobics could climb up the short stairway to the cabin's flat roof, where chairs were set out and, weather permitting, fresh air and the passing views made for pleasant relief from the congested cabin.

Yet finding space to settle down was no minor matter. The cantankerous Frances Trollope provides a vivid picture of what this contest was like: "In such trying moments as that of fixing themselves on board a packetboat, the men are prompt, determined, and will compromise any body's convenience, except their own. The women are doggedly steadfast in their will, and . . . look like hedgehogs, with every quill raised, and firmly set, as if to forbid the approach of any one who might wish to rub them down. . . . Even the youngest and the prettiest can set their lips, and knit their brows, and look as hard and unsocial as their grandmothers."[11] Mrs. Trollope, it must be admitted, took a bitter view of just about everything she encountered in America. Here, for example, is what she had to say about the town of Avon, south of Rochester: "It is a draggling ugly little place and not any of their 'Romes, Carthages, Ithacas, or Athens' ever provoked me by their name so much. This Avon flows sweetly with nothing but whiskey and tobacco juice."[12]

The roof of the packet boat may have provided moments of peace and quiet, but never for long. There were constant interruptions caused by the several hundred bridges connecting farmlands and other properties split apart by the canal's route. These bridges had been built as low as possible in order to facilitate the movement of farm animals or equipment across the canal, leaving perilously little space between the bottom of the bridge and top of the canal boats. At the boatman's all too frequent cry of "Low bridge!" the occupants of a canal boat's roof had to rush downstairs or throw themselves on the flat surface of the roof—no easy task for the ladies clad in voluminous skirts and elaborate hats.

The literature about the canal is full of stories about people who got caught in the process and were either dumped into the water or just barely escaped being squashed. Hawthorne relates what happened when a Virginia schoolmaster, "too intent on a pocket Virgil to heed the helmsman's warning 'Bridge! bridge!' was saluted by the said bridge on his knowledge-box. I had prostrated myself, like a pagan before his idol, but heard the dull leaden sound of the contact, and fully expected to see the treasures of the poor man's cranium scattered about the deck. However . . . there was no harm done, except a large bump on the head, and probably a corresponding dent in the bridge."[13]

* * *

The discomforts of daytime were minor compared with the adventures when the cabin was converted into sleeping quarters at night. One traveler depicted a cabin equipped for thirty men and fifteen women, where the men were confined to a sleeping area twenty-one feet long, leaving some of them without a place to lie down.[14] The "sleeping apparatus" consisted of three tiers of canvas-bottomed frames, "hardly broad enough to allow the occupant to stretch himself on his back," which meant no space to sit up or change one's clothes.

The process of climbing into the sleeping apparatus was more difficult for some than for others, "as those who were blessed with any thing like rotundity of person felt considerable difficulty in getting fairly into the narrow recess." This traveler considered himself better off than that: "Although enjoying less room, I believe, than if I had been a mummy in one of the Pyramids, I passed a very conscious and refreshing night." Yet the space was so crowded and the air so foul that, at dawn, "I crept from my shelf with all the caution of a snail from its shell, for with undue haste my nose might have run foul of some protruding stern quarters, or my toes saluted the gaping mouths of the prostrate snorers."[15] Hawthorne had an ingenious solution to the problem of snorers: "Would it were possible to affix a wind instrument to the nose, and thus make melody of a snore, so that a sleeping lover might serenade his mistress, or a congregation snore a psalm-tune!"[16]

These conditions did not improve with the passage of time. Although a travelers' guide published in 1841 favored the packet boats over the rocking ride on railroad cars or the bumps and swings of stagecoaches, readers were warned that "the lodging part, if there are many passengers, cannot be favorably spoken of."[17] About the same time, a Scotsman observing his fellow passengers suspended in rows in their bunks was reminded of "strings of onions in a green-grocer's shop," and Philip Hone, mayor of New York in 1825, complained in 1835 that his neighbor overhead "packed close upon my stomach."[18] Hone found conditions in one of the Pennsylvania canal boats in 1847 even worse, when he was inspired to compare the passengers in their sleeping quarters to "dead pigs in a Cincinnati pork warehouse."[19]

Not everyone took such a dim view of canal travel. One traveler's italics made clear how much he appreciated the elegance of his packet: his boat had "*three* horses [to] go at a quicker rate and have the preference in going through the locks . . . are built extremely light, and have quite Genteel men for the Captains, and use *silver* plate. . . . The table is set the whole length of the boat [and] is supplied with everything that is necessary and of the best quality with many of the luxuries of life."[20] A young bride in 1827 reported that the breakfast set out for the passengers on her packet boat included "pike or bass . . . steak, bacon, sausage, and ham . . . scrambled eggs, baked pritties, boiled cabbage and squash, bread . . . pancakes . . . with maple or honey to choice; and to wash all down, coffee, tea, milk, skimmaging, and cider." And then she adds that "Dinner will be heartier."[21]

All boats carrying passengers, line boats as well as packets, provided, at the least, a steward, two helmsmen to take turns steering the boats (most of which kept going twenty-four hours a day), two drivers to take turns on the animal teams, and a captain (the crews on the canal boats, both passenger and freight, were known as "canawlers," and for many years "canal" itself was pronounced "canawl").[22] Day and night, one or another of these men or boys would toot the horn or blow the trumpet every time they approached a lock or identified another boat in the opposite direction. With the canal so crowded even in its earliest days, these precautions were extremely important. As an additional safety measure for night travel, the boats carried large reflecting lights on either side of the bow. When a number of boats either were moving along together or passing one another in opposite directions, the impression was like strolling on a brightly lit street rather than being pulled along water through the dark by a team of horses or mules.

* * *

Most people who started or finished their trips in Albany took the stagecoach between Albany and Schenectady rather than consuming the twenty-four hours required to negotiate the twenty-seven locks in a distance of only twenty-four miles. These coaches were so popular that as many as thirty a day made the round trip, with few vacant seats, for a fare of 62¢. Sixty-two cents—about $7.50 in today's money—was a lot more than it appears to us. A captain of a packet boat earned $30 a month, or about a dollar a day, and the driver of the horse or mule team received only $7 to $10 a month; at this date, members of the canal commission earned $2500 a year, or less than $10 a working day (the president of the United States fared better at $25,000 a year).[23] The 62¢ seems even more costly compared with the Greyhound bus fare from Albany to Schenectady in mid-2003, which was $4.50 on weekdays, approximately 10 percent of a day's pay at the minimum wage and far smaller than that for the vast majority of travelers.

Fares on the packet boats varied with the pretenses to elegance but they averaged around 4.4¢ a mile, which included meals; passengers choosing to pass up the meals paid about a penny a mile less.* The 152-mile trip from Utica to Rochester, a matter of two days and two nights, cost $6.25 and the 86 miles between Schenectady and Utica cost $3.50; a

*The relatively luxurious steamboat from New York to Albany, a trip of 150 miles, cost $3 without meals, or 2¢ a mile. Other passengers mentioned a $2 fare. See Warren Tryon, *A Mirror for Americans*, vol. 1, pp. 105–6.

line boat over the same distance charged an even four cents per mile, although in larger towns passengers often went from one boat to another bargaining with the captains to find the best deal.* One American traveler reported paying only 1.25¢ a mile for the Schenectady-Utica trip on "a very superior boat . . . and no charge for births [*sic*], which are a very necessary accommodation . . . and never was I more delightfully situated as a traveller than on this occasion."[24]

The packet and line boats changed mule or horse teams every fifteen to twenty miles, day and night, but the freight boats typically had their own stables located in the bows, with one team of animals working while the other was rested and fed. The canal itself became the repository for both human and animal waste.† No wonder there were complaints about bowels in a prodigious uproar.

* * *

Maintenance along the route of the canal was a constant concern, although the artificial river of the Erie Canal was a lot more manageable than the Mohawk River had been for the Western Inland Lock Navigation Company in the 1790s. Heavy rains, especially in the spring, often caused the canal to overrun its banks and the towpath, requiring crews to hurry out and make repairs. On occasion, they rushed to build a dam to hold the water back while repairs were under way. When that happened, the portages of the old days reappeared, with wagons or bare hands carrying freight and passengers around to the next navigable spot. Boat traffic backed up at the same time, resulting in costly delays.

Unlike Washington's Patowmack Canal, the Erie Canal froze solid in the winter months and all travel and work came to a halt—tempting the canawlers to get just drunk enough to be hauled to jail, where they could

*Freight boats charged 5¢ per ton-mile for merchandise and 3¢ for produce (Ronald Shaw, *Erie Water West*, pp. 200–201).

†Today's travelers on the River Thames in England would be horrified. The law there stipulates that nothing, but nothing, can be thrown overboard, and special "sanitary stations" are located at the locks for the disposal of human waste.

live rent-free until the spring.[25] When the winter broke, an immense amount of work was necessary before the canal could be brought back into operational condition and the boats and the crews could emerge from hibernation. The repairmen would find long stretches where the cold, ice, and underwater currents of winter had disrupted the shape of the bottom of the canal, with huge bumps blocking easy passage by the canal boats. Snow and ice also distorted the banks, and much of the towpath area had to be rebuilt and set solid.

Despite these obstacles, Yankee ingenuity over time managed to achieve significant improvements in the operation of the canal. By 1846, the average weight of cargoes was 66 percent greater than ten years earlier.[26]

* * *

When the great French economist Michael Chevalier visited the Erie Canal in 1835, he could declare without qualification, "The results of this work have surpassed all expectations."[27] The first ten years after the completion of the canal in 1825 had transformed the landscape of the United States all the way from New York City to the budding metropolis of Chicago eight hundred miles away on Lake Michigan. Villages were becoming towns and towns were becoming cities. Rushing rivers and steam engines were powering factories, and by 1831 the railroads would come along to supplement the Erie Canal in moving people and cargoes. Indeed, a major enlargement of the canal itself would soon be on the drawing boards.

Chevalier was much more interested in trade and commerce than in quality of life. But in the early years of the Erie Canal, there was the same kind of tension between industry and quality of life as in our own day. In 1820, the English tourist John Howison had sensed what was coming when he asked, "How long was this delightful state of things to last?"

In 1828, a Yankee had provided Fanny Trollope with a prescient answer to Howison's question. After apologizing to Mrs. Trollope for the wild state of the country they were passing through, this man went on to point out that the property all round thereabouts had been owned by an Englishman, "And you'll excuse me, ma'am, but when the English gets a spot of wild ground like this here, they have no notions about it like us; but the Englishman have sold it, and if you was to see it five years hence, you would not know it again; I'll engage there will be by that half a score elegant factories—'tis a true shame to let such a privilege of water lie idle."[28]

Nathaniel Hawthorne provided a grimmer response in 1835, at just about the same moment Chevalier was expressing his enthusiasm. Hawthorne recalled one of the boats he had encountered on the canal—"a black and rusty-looking vessel, laden with lumber, salt from Syracuse, or Genesee flour"—and then he provided a black description of what canal construction and those wondrous stump-removers had done to the virgin forests:

> The forest which covers it, consisting chiefly [of] trees that live in excessive moisture, is now decayed and death-struck, by the partial draining of the swamp into the great ditch of the canal. . . . In spots, where destruction had been riotous, the lanterns showed perhaps a hundred trunks, erect, half overthrown, extended along the ground, resting on their shattered limbs, or tossing them desperately into the darkness, but all of one ashy-white, all naked together, in desolate confusion. . . . The scene was ghost-like—the very land of unsubstantial things, whither dreams might betake themselves, when they quit the slumberer's brain.[29]

In the same vein, Captain Basil Hall, an English tourist, complained that Americans can hardly conceive the horror of a foreigner coming upon "such magnificent trees standing around him with their throats cut, the very Banquos of the murdered forest."[30]

These observations, and many like them, appear to anticipate the ecological nightmares of our own time. But the bias here runs deep.

Descriptions of this kind ignore a process that has gone on since the beginning of civilization: the destruction of the forest to make way for farmland, villages, and ultimately towns, cities, industrial complexes, roads, railroads—and canals. What happened in New York State to build the Erie Canal was nothing new at all, even if its magnitude was imposing.

Furthermore, strong positive social forces were also at work along the canal. Attendance at schools and the level of education was consistently higher in the canal counties than in the rest of New York State. According to one authority, the number of New Yorkers going to college in 1829 exceeded the number from New Jersey and Pennsylvania combined.[31]

At the same time, the canal's revolutionary impact on the economy and the networking of a new nation in an expansionary mode shattered social structures as well as the ecology of the countryside. Here, too, the results surpassed expectations, but not always in a favorable direction. What psychologists like to describe as a sense of alienation took hold and soon affected large numbers of people in the area.

Although the New York economy remained primarily agricultural until well after the Civil War, the rapid growth of commerce and industry uprooted patterns reaching back two centuries. On the farm, the family had always been the center of economic activity. Now the traditional family structure began to crumble as men went to work in the new factories, mills, stores, and offices, leaving their women at home. Or the women, and occasionally the children, also went out to work as the manufacture of textiles moved from the spinning wheel at home to the mill-powered factories along the route of the Erie Canal.[32] By the 1830s, children accounted for more than a quarter of the workforce on the canal itself, working long hours as drivers who handled the horses and mules or doing menial tasks on the boats, like scrubbing the decks and keeping the lines unsnarled.[33] A resident of the town of Watervliet near Albany—the original home of the Shaker religion—testified in 1839 to the canal board that "the Boys who Drive the horses I think I may safely say that these boys are the most profain [*sic*] beings that now exist on the face of this whole erth without exception."[34]

The canal towns developed into veritable seaports, with piers, cargo forwarders, mercantile establishments, and a profusion of grog shops; one report counted more than fifteen hundred of these enterprises along the banks of the canal.[35] Although the advertising by boat companies emphasized the sobriety and good behavior of their crews, a dark and boisterous side of life was soon flourishing among the canawlers and in the taverns. In 1831, the *Rochester Observer* proposed that the canal should be renamed "the Big Ditch of Iniquity."[36]

Herman Melville was so horrified by the behavior of the canawlers in the 1840s that he devoted a long passage to it in *Moby-Dick*, with Captain Ahab protesting that, "Through all the wide contrasting scenery of those noble Mohawk counties; and especially, by rows of snow-white chapels . . . flows one continual stream of Venetianly corrupt and often lawless life. There's your true Ashantee, gentlemen; there howl your pagans."[37] Was it any wonder that a New York State report of about the same date could declare that a quarter of the inmates at Auburn prison had "followed the canals."[38]

* * *

This environment, and the whole sense of something brand-new and unproven, would in time provoke extreme responses among religious fanatics, reformers, and would-be social workers. The stretch of New York from the Hudson to Lake Erie came to be known as the Burned-Over District, combining images of "the fires of the forest and those of the spirit."[39]

The most puritanical groups active in the Burned-Over District drew their numbers from the Yankee immigrants, the less educated and more restless New Englanders who came primarily from the western side of the region rather than the coastal areas. They began a fervid campaign for temperance, which reached the point where as distinguished a member of the community as Eliphalet Nott, president of Union College in Schenectady, could declare that alcohol in the stomach was vulnerable to spontaneous combustion that would burn the drinker to ashes.[40]

Heated controversies broke out over what was called sabbatarianism, or the movement to prohibit traffic movement on the Erie Canal on Sundays. The American Seaman's Friend Society took a practical view of the importance of temperance and Sunday closings by pointing out that a sinful set of crews on the canal would be bad for business. Without "steady, honest, and respectable persons, of both sexes," working on the canal, they argued, "the dregs of the community" would be working there and no respectable citizen would want to make use of it.[41] The opponents of sabbatarianism took the exact opposite position by contending that Sunday closing would back up boats all along the length of the canal, and, even worse, it would be a prescription for the canawlers to take advantage of twenty-four hours of idleness to get drunk and stay drunk, and goodness only knows what other depravity might take over to fill the idle hours. The canal remained open around the clock despite the brouhaha, which lasted well into the 1840s, even though a few lines sought customers by boasting of their closing down on Sundays.

* * *

All of these changes coincided with what the historians describe as the Second Great Awakening—a powerful wave of revivalism and evangelism aiming to hasten the Second Coming of Christ. There was a plethora of fiery sermons, camp meetings attended by thousands, and a sudden increase in church attendance in the urban centers. Although this development provokes many patronizing comments in the literature on the canal, in fact it was testimony to the powerful democratic spirit in the area. The evangelists talked to the people and were of the people, stimulating rapid growth in the Baptist and Methodist churches while attendance declined at the older and more hierarchal congregations, such as the Congregational church.[42]

The coining of the term Burned-Over District for the Erie Canal area is sometimes ascribed to the most zealous, and probably most successful, of the evangelists, Charles Finney. Finney was a fiery preacher who gathered huge crowds to hear him preach virtue with a capital V. Finney underscored the urgency of carrying the moral word to the thousands of dissolutes living and working along the banks of the Erie Canal. He spent about ten years in the area but then, in the summer of 1835—despite realizing that he would greatly miss the excitement of preaching to his large and responsive audiences—Finney became a theology professor at Oberlin College in Ohio. He accepted the Oberlin offer only under the stipulation that the faculty should be allowed to receive "colored people" on the same conditions as whites and that discrimination of any kind on account of race would be prohibited. Oberlin, then just two years old, was the first college in America to accept African-American students and would also become the first to go coeducational.

The most famous of the religious leaders associated with Erie Canal country was Joseph Smith, who inaugurated the Church of Jesus Christ of Latter-day Saints, or the Mormons. Smith was a Vermonter, born in Sharon in 1805 to a family who regularly saw visions, heard heavenly voices, and believed in miraculous cures. When he was about ten years old, the Smith family moved from Vermont to Palmyra, on the canal about twenty-five miles southeast of Rochester.

Although he was given to both visions and epilepsy as a child, Smith was able to recover from the illness and became a crystal gazer with a "peepstone" to help him and his father discover hidden treasures. On September 21, 1823, he claimed later, the angel Moroni appeared to him three times to tell him that the Bible of the western continent was buried nearby, at a spot now known as Mormon Hill. Four years after that, Smith reported he had dug up a stone box containing his "gold bible." As he could read but did not know how to write, he dictated the Book of

Mormon to his wife and several associates in 1830, which was then printed in an edition of five thousand copies. Not long after, the Church of Jesus Christ was formally established nearby at Fayette. Persecution began almost at once, driving Smith and his group out of New York and across the border into Ohio.

More intense and persistent persecution attacked the Catholics, whose numbers increased as German and Irish immigrants continued to move into western New York. As immigrants, the Catholics were often willing to work at wages lower than the amounts earned by men born in America. In 1834, Samuel F. B. Morse, the inventor and painter, published a series of letters in the press claiming that the pope, in cooperation with unnamed European powers, was planning to subvert American society by encouraging immigration.[43] All this opposition accomplished little: at the outbreak of the Civil War in 1860, Catholicism was the single largest denomination in New York and in the nation at large.

Bigotry struck at the center of New York State politics in September 1826, when William Morgan, a Mason originally from Rochester, published in the *Batavia Advocate* what he claimed were the secret rituals of the order. This betrayal provoked the local Masons to attempt to set fire to the publisher's shop and then to have Morgan arrested on trumped-up charges. Although Morgan was soon released from jail, he was kidnapped as he left the prison and was never seen again; it was rumored that he had been tied to weighted cable and dropped into the Niagara River where it joins Lake Ontario.[44]

Because of widespread convictions that Masons practiced secret, evil rituals at their meetings—although the Masons insist that their so-called secrets are solely used as a ceremonial way of demonstrating that one is a Freemason—anti-Masonic movements had always been present to some degree all over the United States. Now the tendencies toward persecution were fueled by the intense emotional environment of the Burned-Over District. Crowds all across the state broke out in wild and noisy protests demanding the destruction of the Masonic order. These movements became veritable crusades and soon turned into anti-Masonic political organizations, claiming that Masonic influences dominated and were corrupting both local governments and the state legislature. Charles Finney later asserted that the hysteria forced just about all the lodges in western New York to shut down.

Two years later, anti-Masonic political activity in New York centered on Andrew Jackson's membership in the Masons and the powerful Van Buren party's support for Jackson's run for the presidency. In an odd twist to this story, Van Buren now found himself on the same side as his old nemesis De Witt Clinton, but Clinton did not live to see the results of the forthcoming contest.

Jackson succeeded in the presidential election of 1828, and his ally Van Buren was elected governor of New York. Jackson squeaked through with less than 51 percent of the popular vote nationally, while Van Buren earned three thousand fewer votes than the combined votes of his two opponents. As Jackson and his faithful New York ally had been very popular before the excitement erupted over Morgan's disappearance, these election results in calmer times would in all likelihood have turned out to be much more one-sided victories for both men.

* * *

None of the canal's most enthusiastic advocates, nor any of its more skeptical supporters, ever promised that a channel of high virtue and social stability would be constructed across New York State. The primary vision was Elkanah Watson's prediction in 1791 of "a water communication of several thousand miles [to] be opened from the Atlantic, to the most extensive inland navigation to be found in any other part of the world. . . . Suffice it to say, the [result] would be more precious than if we had encompassed the [Bolivian gold] mines of Potosí."[45]

The key to Watson's vision was in its final words. In the end, the Erie Canal was about creating a network that would be the equivalent of a gold mine for people seeking to earn great profits from business, money, trade, and industry. As the French foreign minister had observed about Americans as far back as 1785, "These people have a terrible mania for commerce."[46] Most Americans shared George Washington's concern about uniting a nation, but even he had emphasized the importance of trade and commerce in achieving that vital objective.

On that score, the canal delivered everything that could be hoped for. As Hawthorne had predicted, the route of the Erie Canal would indeed turn into "one thronged street, from Buffalo to Albany."

CHAPTER 19

THE PRODIGIOUS ARTERY

In his Memorial of 1816, De Witt Clinton had predicted that the Erie Canal would "convey more riches on its waters than any other canal in the world," releasing resources "to be expended in great public improvements; in encouraging the arts and sciences; in patronizing the operations of industry; in fostering the inventions of genius, and in diffusing the blessing of knowledge." Just nine years later, this far-reaching vision was about to become a reality.

The Erie Canal would transform New York into the Empire State, standing on what Clinton had portrayed as "this exalted eminence."* The national impact of the canal was even greater. The dramatic reduction in travel time on an east-west route into the heartland of the country was an imperative for building a great nation across a huge and fertile continent, where trade, money, and business were rapidly becoming second nature. This narrow ribbon of ditch, less than 375 miles long, provided the spark, the flashpoint, and the inspiration for a burst of progress in America that would eventually coin the buzzwords of the early twenty-first century: economic growth,

*As mentioned earlier, this expression has no clear provenance, although it appears to have been in general use by 1825.

urbanization, national unity, globalization, networking, and technological innovation.

It was no coincidence that the Erie Canal inspired the route of the very first steam railroad in the United States, the Mohawk & Albany, which opened for business in 1831 between Albany and Schenectady, pulled by a locomotive sporting the name *De Witt Clinton* on its coal car. Although only sixteen miles long, the Mohawk & Albany took passengers in one hour past the twenty-seven locks and a full day's travel needed to cover this distance by the canal.

In time, the railroads would eclipse the Erie Canal and the complex network of canals it inspired across the country from Chicago to the eastern seaboard, but it would be a long time. Steam mattered more in powering the boats on the Great Lakes, which brought population to the lake ports and enhanced the connection between the Erie Canal and the lands farther west. And well before the railroads could make a difference, a major innovation in networking would arrive: Samuel F. B. Morse's telegraph lines, which began to weave their way across the nation in 1844, along the Erie Canal and along the new railroad lines as well. The flood of information pouring at breathtaking speed across those telegraph wires radically collapsed both time and distance.

However, as late as 1852, thirteen times more freight tonnage was carried on an enlarged Erie Canal than on all the railroads in New York State.[1] This huge disparity reflects the mix of business in the early years of the railroads, which was only incidentally to carry merchandise and primarily to transport passengers under more comfortable conditions than the crowded and often disagreeable conditions aboard the packets. The railroad system in its early days was not sturdy enough to carry the heavy bulk of grain and timber that sailed with so little effort on the waters of the Erie Canal.[2]

Consequently, the canal continued to put up stiff competition even as the railroads matured. When tolls were abolished in 1882, the Erie Canal was serving over twenty million people annually and had produced revenues of $121 million since 1825, more than quadruple its operating costs.[3] And it was still going strong.

After a significant enlargement in the 1840s, the canal went through a second and far more impressive enlargement at the turn of the century, when total freight traffic exceeded six million tons—triple the volume in 1860.* The moving spirit in this massive project was none other than George Clinton, grandson of De Witt Clinton and the namesake of De Witt's beloved uncle. This George earned the title of "Father of the Barge Canal" for his contribution.[4]

Once again, New York State turned to the federal government for financing, as it had at the very beginning of the Erie Canal, and once again was disappointed. Governor Theodore Roosevelt was not to be deterred, proclaiming, "We [New Yorkers] cannot afford to rest idle while our commerce is taken away from us, and we must act in the broadest and most liberal spirit if we wish to retain the State's supremacy. . . . While giving all weight to the expense involved, we should not be deterred from any expenditure that will hold the supremacy of which we are all justly proud."[5]

Financed by a bond issue of $101 million—the equivalent of a billion dollars in today's money—digging on the enlarged Erie began in 1905 and continued for thirteen years. This was more than fifteen times the cost of the original Erie Canal. It was also almost double the time required for its construction from 1817 to 1825, even though very powerful steam shovels and explosives had long since replaced the hand shovels and blasting powder of 1817. When completed, the new canal could carry huge barges with a draft of 10 feet, 250 feet long, and 25 feet wide, towed by steam-powered tugboats on an uninterrupted waterway between the Atlantic Ocean and Buffalo, leading to the Great Lakes and the Midwest.†

*For an excellent map showing both the expansion of the Erie Canal and the proliferation of canals in the post-Erie craze, see Jeremy Atack and Peter Passell, *A New Economic View of American History from Colonial Times to 1940*, p. 151.
†Only with the advent of this much deeper and wider barge canal did the Erie Canal commissioners finally abandon mule power along the towpaths.

By 1951, annual freight traffic on the barge canal had risen to 5.2 million tons. Today, the railroad is a much less significant competitor than the St. Lawrence Seaway as well as the New York State Thruway, which extends like a continuous ribbon from New York to Buffalo. Business on the barge canal has dwindled to almost nothing other than pleasure boaters and an occasional freight shipment of modest size.

* * *

But when the Erie Canal opened for business in 1825, it was ready to play a critically important role in the economic development of what was, in today's parlance, still a clearly underdeveloped country. The canal provided a fantastic wealth-creation machine for the powerful forces of economic change at work in the United States, motivated by the unquenchable passion of Americans for money and business and by their impatience to get ahead. This set of attitudes was well summed up by the play on words *Laissez nous faire!* (let's go).[6]

After an American visit in the mid-1830s (one that sounds remarkably like a visit to the United States at the beginning of the twenty-first century), the French economist Michael Chevalier observed that work in this country "goes on *à l'américaine*, that is to say, rapidly. . . . Here all is circulation, motion, and boiling agitation. Experiment follows experiment; enterprise follows enterprise."[7] Like most foreign visitors of the era, he took a trip on the Erie Canal—"simple as a work of art, prodigious as an economic artery"—and was struck by the restlessness of Americans he encountered there. "The full-blooded American," he reports, "has this in common with the Tartar, that is he is *encamped*, not established, on the soil he treads upon."[8]

Chevalier was fascinated by the American devotion to money as the standard, the aspiration, the prime metric of everything, a view he considered alien to his countrymen's view of money. The Americans are not only determined to accumulate money, he pointed out, they consider everything convertible into money: "The maxim here is that everything has to be paid for. . . . All that [the American] has, all that he sees, is merchandise in his eyes. . . . To him a waterfall is simply water power for his machinery. . . . The Yankee will sell his father's house, like old clothes, old rags." He cites Talleyrand, who said, "I do not know an American who has not sold his horse or his dog." Even crimes that would result in prison in France can be satisfied with a fine payable in money. Most astonishing to Chevalier, accused murderers and arsonists in America are allowed to go free pending trial, just as long as they can put up enough money to cover bail.[9] But Chevalier also makes the remarkable assertion that "there are no poor people here, at least not in the Northern and Western States."[10]

Occasionally, Americans saw themselves as others saw them. In 1836, at the cusp of an extended business boom, an American metals dealer wrote home from London that the English "think you are all quite wild in America." And Henry Remsen, former president of the Bank of the Manhattan Company and a professional speculator, observed that Americans were "apt to do too much business [in their drive to achieve] splendor and idleness."[11]

* * *

We have crude but impressive numbers to describe the impact of the Erie Canal on American economic development. Estimates of real gross national product—the total output of goods and services after adjustment for a changing price level—show that economic growth from 1800 to the opening of the canal in 1825 ran about 2.8 percent a year, something similar to the long-term trend in economic growth in our own time. But as a consequence of both the War of 1812 and the boom and bust of 1819–1821, there was a lot of variation around that average. Conditions changed dramatically over the quarter century following the Wedding of the Waters. The growth rate jumped dramatically, to the very high number of 4.6 percent a year, with a far lower degree of variability; this meant that total U.S. output tripled over those twenty-five years—among the highest periods of economic growth in two centuries of American economic history.

These are statistical estimates, but nothing depicts the story of economic change more vividly than the movement of Americans into urban centers. In 1820, only 7 percent of the nation's population could be classified as urban, and even that proportion was only 2 percentage points higher than it had been in 1790. But five years after the opening of the Erie Canal, in 1830, 9 percent of the country was living in urban centers, and this number jumped to 15 percent by 1850 and continued upward from there.[12]

Much of this change was occurring at the extremities of the canal, at Buffalo and Albany, and then down the Hudson to the dazzling port city of New York, with its harbor open to the lands across the seas. In 1789, Elkanah Watson was shocked at Albany's unpaved streets and lack of street lamps, and in 1831, as we noted earlier, Alexis de Tocqueville described his approach to Albany as "tilled fields [with] trunks in the middle of the corn. Nature vigorous and savage." But the population of Albany County doubled to 28,000 over the ten years after the opening of the canal in 1825; by 1850 it was up to about 90,000 people. Five years after the opening of the canal, the old city had become just the core of a much larger community of roughly one hundred new blocks.[13]

Kingston, a tiny village on the Hudson River before the completion of the canal, developed into a small city for shipping coal received from Pennsylvania but also manufacturing cement, machinery, and iron castings. This was just one step in converting the Hudson valley into an industrial network that included other river towns such as Newburgh, Poughkeepsie, Hudson, and Troy. All of these towns shipped a rising volume of merchandise westward through the Erie Canal while the trade coming to them from the other direction reached magnitudes they never would have thought possible fifteen years earlier.[14]

At the canal's western terminus, Buffalo provided an even more spectacular performance. A village that De Witt Clinton had scorned in 1810 for having "five lawyers and no church," and where a later visitor's bowels had been set "in an uproar prodigiously," expanded from a population of 2600 in 1824 to 15,600 ten years later and 42,000 in 1850. In 1830, the *Buffalo Journal* could boast that "our children are surrounded by the comforts, the blessings and the elegances of life, where their fathers found only hardship, privation, and want."[15] In 1831, de Tocqueville, not easily pleased by what he saw in America, commented on the pretty shops and French goods he noted on a walk through town (although he found "not one Indian woman passable").[16]

At the eastern end of the canal system, New York City was one of the major metropolitan centers of the world by 1850, with a population of 700,000 people, quadruple the number when the Erie Canal opened for business in 1825. Over the eleven years from 1825 to 1836, the value of real and personal property in New York City tripled, from $101 million to $310 million—and from 38 percent to 48 percent of the total for New York State—while the population was increasing by "only" 63 percent.[17]

Meanwhile, a lot was happening along the canal between Albany and Buffalo, succinctly summed up by the rapid appearance of ten brand-new towns located between Syracuse and Buffalo and readily identified by names ending in "port"—which signified extended docking and loading facilities along the canalside. When canal construction began in 1817, Rochester and Syracuse together had fewer than 3000 people and Lockport was just making its appearance on the scene. But by 1825, the three towns had a total of about 6000 residents and by 1850 their population was twelve times that. As a result of this pace of development, land values along the route of the Erie Canal grew by 91 percent from 1820 to 1846.[18]

* * *

In Britain, from the 1760s to about 1800, a great canal boom had made the Midlands the center of the Industrial Revolution, attracting population, capital, and entrepreneurs from all corners of the kingdom.* Merrie Olde England—indeed, much of the rest of the world—was

*These were the canals Canvass White had visited in 1817.

never the same again. In the same fashion, the networking effect of the Erie Canal extended far beyond the borders of New York State. Although the canal terminated at the eastern end of Lake Erie, its influence spread rapidly westward as it carried almost all the people moving out to Ohio and then on to the Great Lakes. The population of the west rose from about 2.5 million when the Erie Canal was completed in 1825 to 7.5 million in 1850, or from 21 percent to 33 percent of the country's total population. Without that water route through the Allegheny Mountains, George Washington's fears that "the touch of a feather would turn them away" would in all likelihood have become a reality. Under the best of circumstances, the movement of people into the western lands would have progressed far more slowly, delaying the conversion of those lands into the breadbasket, and ultimately the industrial center, of the United States.[19]

In the process, the canal changed the face of the nation by transforming its primary axis from north-south to east-west, and away from George Washington's Potomac route to the mountains, with momentous consequences for its future history. As the flood of New Englanders and European immigrants moved west on the Erie Canal packets, the momentum of America shifted increasingly from the slaveholding and cotton-producing South to the free labor and Industrial Revolution of the north.[20] Indeed, no comparable network of canals existed in the South—a disparity that helps to explain northern superiority in the Civil War that came two decades later.

The rapid expansion of the towns and cities along the canal is only a partial view of the impact of the westward flow. The opening of the canal also led to an emphatic slowdown of growth in rural areas along the waterway at the same time that the urban communities like New York, Rochester, and Buffalo were flourishing. The canal passed through sixteen counties, from Albany County on the east to Erie County on the west; in 1820, five years before the completion of canal construction, these counties accounted for 38 percent of New York State's population.

By 1850, their share was down to 28 percent. Their population had grown by only 60 percent over those thirty years, while those of both New York State as a whole and the nation were more than doubling.

What happened? The process did not go forward in a straight line, but the outcome appears to have been inevitable. In the early years of the canal, agricultural activity boomed along its route, as the area's rich production of wheat, barley, oats, and corn found an artery of transportation to feed the east. This development explains the rapid growth of Rochester in particular, which was the center of the grain-producing area and provided north-south transportation along the Genesee as well as its facilities on the Erie Canal. Genesee flour, famous for its "sweetness and fineness," accounted for the largest share of the Erie Canal's eastbound freight until the mid-1840s.[21] But over time, as the canal facilitated growth in the west, and as the fabulous fertility of those lands became increasingly apparent, many New York farmers caught the fever and joined the crowds of people from other states heading farther west to seek their fortunes.

Indeed, the canal attracted people to the new territories from all over the eastern states. Vermont and New Hampshire were especially affected by these forces, as the population of both states stagnated between 1830 and 1850. The easterners who stayed home had to be philosophical about this powerful shift from east to west and made the best of it by recognizing the canal's strong influence on the unification of the nation on both sides of the mountains. As Levi Woodbury, former governor of New Hampshire, wrote in January 1834 from Vermont to one of the Erie Canal commissioners, "I can almost submit to be envious. But [after] a moment's consideration, that you open your generous arms to welcome the emigrants from our frosty hills & to patronize [our] sons. . . . I feel again, friend, that we are in many respects *but one people* and that the success of a part is in some degree the success of the whole."[22]

Canadians, who were similarly affected, took a less cheerful view of the impact of the Erie Canal. As Elkanah Watson had predicted back in 1788, "The state of New-York have it within their power, by a grand stroke of policy, to divert the future trade of Lake Ontario and the great lakes above, from Alexandria [in Ontario] and Quebec, to Albany and New-York."[23] And they did. The canal diverted so much traffic that Canada was feverishly building their own canals, notably the Welland Canal to bypass Niagara Falls and connect its side of Lake Ontario to Lake Erie. The Americans were not about to accept the competition without a response. They dropped their duties on Canadian merchandise and lured the traffic back to the Erie Canal. The result was a serious loss of Atlantic traffic moving through Montreal, with accompanying bankruptcies and rising unemployment.

With all this going on, change was both rapid and pervasive in the territories to the west of New York State. Just as an example, the state of Ohio, with its fertile fields, long coastline on Lake Erie, and river connection to the Mississippi, would more than triple its population over the twenty-five years after the opening of the Erie Canal, starting from an already respectable 581,000 in 1820. This growth advanced Ohio from less than 2 percent of the total U.S. population to more than 4 percent.

* * *

Along the way, the Erie Canal performed an entirely unanticipated and unrelated function, one that developed into one of its major activities during the 1830s. The state had reported a surplus of nearly $600,000 in the Canal Fund for the year 1829, setting in motion a wave of revenue well in excess of operating costs and the interest payments due on the outstanding debt. As a result, the commissioners began to spread the good fortune around New York's commercial banking system, some in the form of deposits but also in the form of loans. They also made efforts to buy back in their 6 percent bonds due in 1837, but the holders of those bonds, most of whom were foreigners, were too pleased with their investment to part with any paper, except at prohibitively high prices, before the due date finally arrived.* This whole process involved the commissioners in the economic development and the financial network of the state, and many localities beyond New York's borders. In the aftermath of the shattering financial and economic crash of 1837, the Canal Fund even performed like a modern central bank by using its surplus of revenue to serve as a lender of last resort to many liquidity-strained banks in New York State.[24]

As happened centuries earlier along the roads the Romans built across all of Europe and along the famous Silk Road to the spices of the Far East, some of America's greatest cities came into existence along the new east-west axis. In 1820, fewer than 100 people were living on the future site of Cleveland, Ohio, but within a few weeks of the opening of the Erie Canal, fifteen vessels sailed from Buffalo with merchandise from New York for Ohio. A year later, the first shipment of pork left Ohio for the seaboard, and as the years went by pork became one of Ohio's largest eastbound exports, later to be supplemented by both whiskey and potash. By 1830, Cleveland's population was up to 1000 and there were 17,000 Clevelanders by 1850. The growth spurred by the canal from the east toward Chicago was even more dramatic. Connected early on by the railroads as well as by water to all four directions of the compass, the population of Chicago reached 4500 in 1840 and 30,000 ten years later, after starting at about 100 in 1830.

By that time, Chicago had its own canal connection, the ninety-six-mile-long Illinois and Michigan Canal, which connected Chicago and the Great Lakes southwestward to the Illinois River and from there down to St. Louis and the Mississippi. Built in the canal mania that followed the completion of the Erie Canal, and finished after ten years of work

*This is an oversimplification. The Erie Canal bonds were selling at a premium—a price over par, or 100—both because of their high quality but also because the interest rate they paid was higher than going rates in the first half of the 1830s. That premium would disappear on the redemption date in 1837, when the bonds were be paid off at par. Consequently, some holders did sell their bonds back to the Canal Fund a short time prior to redemption.

and the default of the State of Illinois on the bonds that financed it, the canal opened an additional route for the output of the midwest and the cotton in the Mississippi delta to move out to the eastern seaboard at New York City and the Atlantic Ocean.[25] Chicago was also just the beginning of another and even greater network. In 1835, fewer than 5000 white people lived in the huge expanse between Lake Michigan and the Pacific Ocean, but twenty years later more than 1 million had settled out there.

As population exploded in the western areas, the huge volume of raw materials in the midwest—grain and flour above all, but timber and coal as well—was liberated by the Erie Canal from the circuitous and treacherous route down the Mississippi River to New Orleans and the Gulf of Mexico. Indeed, insurance to cover a shipment over the 1600 miles from New York to New Orleans by way of St. Louis and the Mississippi was more costly than insuring a shipment traveling the 3000 miles from New York over the ocean to England or France.[26] The natural wealth of Pennsylvania, Ohio, Indiana, and Illinois could now move in bulk directly to the big commercial centers and ports like New York and Boston on the Atlantic coast and beyond them to the great markets of Europe. Over time, new canals in all these states, and the railroads after the 1840s, created an even larger and more elaborate transportation complex to augment the capacity of the Erie Canal, accelerating the movement of both heavy and light freight to the Atlantic coast and of passenger traffic heading west for a new life on the other side of the mountains.

None of this was instantaneous. As noted, it took time for the combination of a growing population beyond the mountains and an increasingly complex transportation system to work its miracles. At first, the level of eastbound commercial traffic moving toward New York from the west was much smaller than the merchandise of all kinds heading westbound. But once under way, the eastward flow of freight rapidly gained momentum. In the mid-1830s, the traffic moving toward New York was still low at around 50,000 tons, only one-sixth of the tonnage moving toward the west. But fifteen years later the eastbound trickle had turned into a flood. In 1847, the tonnage of grain, meat, dairy products, and "domestic spirits" (whiskey) exceeded westbound volume for the first time, and it kept on growing.[27]

The transition had been a complex one. First, the Hudson valley was replaced by western New York as the area's main source of grain and flour. Then western New Yorkers succumbed to the wondrously fertile lands beyond the borders of New York, in Ohio, Indiana, Kentucky, Tennessee, and Illinois. As a result, farm productivity increased more than 30 percent between 1800 and 1840. Without the efficient and low-cost transportation facilities provided by the Erie Canal, this spectacular achievement would have happened much later or much more slowly.[28]

By 1850, the volume of agricultural production moving eastward from the lands to the west had replaced nearly all the old sources of grains and flour in New York State (Rochester and the Genesee valley were strong survivors), which now turned instead to dairy farming and cheese, cattle raising, and growing fruits and fresh vegetables for nearby urban areas.* The tonnage of flour and grain reaching Buffalo from the west in the mid-1840s was ten times the volume just a decade earlier.[29] By the time the Civil War broke out in 1861, and even in the face of intensifying competition from the railroads, Erie Canal freight heading for New York from the west had reached 2 million tons annually, while fewer than 400,000 tons were moving in the opposite direction.

* * *

The spirit of change, speed, and innovation was energized along the entire route of the Erie Canal and the new localities connected to it by later canals and the railroads. Entrepreneurial activity was exploding. As

*This was not an easy choice if an unavoidable one. These forms of farming are much more labor intensive than growing grains.

a German visitor described his trip to America as late as 1857, "Ten years in America are like a century elsewhere."[30]

This view was echoed by a British parliamentary commission that visited the United States in the 1850s. The group was especially impressed with how American workers accepted progress instead of resisting it as they "hailed with satisfaction all mechanical improvements [that were] releasing them from the drudgery of unskilled labour, [which] they are enabled by education to understand and appreciate."[31] Coming from a British group, this was an exceptionally high compliment. But the commission went on to emphasize the contrast with their own labor force, which stubbornly resisted "mechanical improvements" for fear of losing their jobs.

The Erie Canal itself made a significant contribution to the pace of technological innovation. The most impressive and reliable evidence of the developing Industrial Revolution is in the growth in the number of patents—the legal foundation of innovation and economic development. In a detailed study of new patents granted by the U.S. Patent Office from 1790 to 1846, Kenneth Sokoloff, an economic historian at the University of California, Los Angeles, points out Patent Office that all the places with high patenting rates were either metropolitan centers or located close to a navigable waterway: "Perhaps the most vivid example is New York, where the completion of the Erie Canal in 1825 seems to have sparked big changes in the composition of output and a sharp rise in patenting along its route."[32]

Change of this nature was revolutionary, in every sense of the word. As long as the movement of human beings and produce—primarily agricultural—was limited to rivers or was ambling along in a trundling and rickety wagon, as it had been for countless centuries, no one thought very much about change. But as a waterway, the Erie Canal immediately increased the expected return to inventors working near a large pool of easily reachable potential customers, competing suppliers, and a growing stream of information.[33]

The impact of the patenting increased with the passage of time. While the Patent Office issued approximately two hundred patents a year from 1810 to 1825, patenting skyrocketed after the opening of the Erie Canal, hitting a pre–Civil War peak of over seven hundred patents in 1835.[34]* New York State led the nation in new patents per capita in almost all sectors of the economy, although southern New England took the lead in manufacturing after 1830—but even then New York remained a close second.

A firsthand report on these trends is left to us from the records of the New York Life Insurance and Trust Company, whose board of directors included such luminaries as John Jacob Astor, Henry Remsen, and the hardware merchant and railroad promoter Erastus Corning. In 1832, the company sent Nicholas Devereaux of Utica to look over business conditions along the route of the canal all the way out to Buffalo. Devereaux's report was unqualifiedly enthusiastic, noting the increase in population, the growth in cities and towns, and the prosperity of the farms. He was especially impressed with the proliferation of flour mills and "manufacturies." Summing up the situation, he could only flatter his superiors for being so "sagacious" in recognizing the high investment returns available in the canal area.[35]

Devereaux ascribed the dynamics of the canal economy to "easy access to market." He had it exactly right. The manufacturing activities that stemmed from patenting and technological innovation needed large and growing markets to justify the investments and risks they involve. The Erie Canal provided the perfect linkages between expanding markets in New York City, Albany, and Buffalo and the territory in between. In time, these linkages spread to the urban centers to the west from Cleveland and Detroit out to Chicago.

All of these developments—the growth of population, the extension of the canal's stimulus to the new lands in the west, and the accelerated

*Kenneth Sokoloff shows all the patent data on a per capita basis to neutralize the rapid growth of population in those years on the volume of new patents.

pace of technological change—combined to create a multitude of new markets large enough to support production sold to buyers elsewhere or consuming products from other markets. Commercialization of economic activity spread rapidly, from simple goods for household use to the manufacture of steam engines and heavy engineering. In contrast, as we have seen, home production of textiles in New York State fell rapidly after the canal opened. Farming, too, was being transformed from "a way of life" to farming as a profit-seeking business.[36] There was an internal dynamic to these trends. As farming ceased feeding just the family and became more of a business, farmers with newfound cash income now joined their urban neighbors as ready customers for the tempting products coming from the factories.

One of the most careful students of the economics of this period, Columbia University historian Carter Goodrich, suggested in his 1960 study of public improvements that the opening of the Erie Canal was the dividing point between the periods of the "Frontier without the Factory" and the "Frontier with the Factory."[37] Manufacturing did score spectacular growth along the canal and, courtesy of the canal, through other parts of New York State. Factories for ironworks, hats, and textiles flourished in Albany, while spinning mules and power looms were churning out textiles in Rochester. There were sixteen textile factories in Oneida County in 1827, employing seven hundred people; by 1832, there were twenty cotton mills, with over two thousand employees. In 1827, the nation's first hardware store, dealing mainly in American merchandise, opened in New York City.[38] The state had 13,667 manufacturing establishments in 1835, with sales totaling $59 million—already an increase of 58 percent since 1814. Just five years later, in 1840, revenues from manufacturing production had jumped by more than 60 percent to $96 million, and they would hit $237 million in 1850.[39]

Average sales of only $4300 per establishment in 1835 sounds tiny to our ears today, even when translated into roughly $43,000 in current purchasing power. But these firms were still single-unit enterprises with only a handful of employees—although they were also becoming increasingly specialized, as Adam Smith would have predicted. In manufacturing, the owner and manager was in most cases a former artisan—such as a blacksmith, tinsmith, or plumber—who was now using advanced equipment for the first time to make products varying from stoves and grates to gas fixtures and steam engines. In trade, middlemen were beginning to operate between the manufacturer and the retailer, or between the importer and the ultimate destination of the goods. These little business firms were formidable economic units for their own time, and it is interesting to note that the Industrial Revolution and technological progress came along a lot faster than changes and innovation in the structure of the business firm.[40]

The influence of a shift to industrialization was especially notable in the cities on the canal. For example, Rochester was famous for its many mills and its dominance over the grain and flour trade, but the expansion in manufacturing was equally impressive. In 1835, a merchant named Henry O'Reilly could proclaim that "the flouring business for which Rochester is at present most celebrated, is by no means of such importance . . . as the other branches of manufactures."[41] But manufacturing would not have developed in Rochester without the Erie Canal's communications system that opened the city to markets large enough to justify the capital investments in factory construction and equipment.

Edward Peck's paper mill flourished in response to the increasing numbers of printers and the expanding market for both newspapers and books. The Cunningham carriage factory was nationally recognized and survived nearly a hundred years, ending its career as an early manufacturer of automobiles. Lewis Selye's machine shop turned out fire engines and railroad cars for other cities and towns. David Barton's Hydraulic Building sold water power to the toolmakers on its upper floors, who in turn supplied the needs of the growing city and the prosperous agricultural communities beyond. Tailors who had begun their careers as craftsmen sewing clothing to order transformed the men's clothing business in Rochester into a mass production industry, which flourished as a source for the entire country into the late twentieth century.*

The canal also transformed Syracuse from a small village into a major industrial center. Situated just about halfway between Albany and Buffalo, the area around Syracuse had been a source of salt since the 1700s. Salt has been one of the most essential raw materials in human history, because of its unique quality of preserving foods and also for curing raw meats such as pork. By opening up a transportation system that could move tons of salt at a time from the Syracuse area, the United States was able to replace the substantial imports of salt from the Turks Islands, Portugal, and the Cape Verde Islands.[42]

The community that was to become the city of Syracuse was originally a couple of houses and a tavern. In 1800, Joshua Forman—the New York State assemblyman who in 1808 played a crucial role in starting legislative action for the Erie Canal—had moved into the vicinity to pursue a law practice. Forman's efforts for the canal ended up paying big dividends for his community, and in 1825, the same year the canal opened, Syracuse was officially incorporated as a village; by no coincidence, its chief executive was Joshua Forman, on hand to officially welcome De Witt Clinton when he came through on the *Seneca Chief* on his way to the Wedding of the Waters.

Originally, salt from the Syracuse area had moved west by pack mules to Lake Erie. But the canal immediately stimulated a steep increase in the demand for salt in the west, not just because of the growing populations of humans but also because of the urgent need for salt in curing the burgeoning supply of pork. The number of hogs to be slaughtered and cured, in turn, was increasing so briskly because they were a market for excess supplies of grain growing so abundantly in the new western lands.

*This clustering process would be immortalized in Alfred Marshall's *Principles of Economics*, the masterwork of the great Victorian English economist, who described it as "the neighborhood effect."

The combination of the Erie Canal and the salt deposits around Syracuse created a large industrial center out of what had been described by an 1820 visitor as "so desolate it would make an owl weep to fly over it."[43] Five years after the opening of the canal, the population of Syracuse had tripled, from 250 to about 750, but that was only the beginning. Once the canal was in operation, salt could be moved in bulk and at low cost to communities to the east as well, and the salt enterprises grew rapidly. By 1850, the population of Syracuse had grown to 22,000—almost all of it thanks to salt (and not incidentally to Joshua Forman). Huge vats and mechanized refineries turned out rising quantities of salt for many uses in addition to food and preservation, including major chemicals such as soda ash, caustic soda, and bicarbonate of soda.

The chemical industry based on salt in Syracuse grew so large it polluted nearby Lake Onondaga to the point of killing off almost all life in its waters. In May 1918, the section of the old Erie Canal running through the city was closed and covered with a boulevard—appropriately named Erie Boulevard; the route of the modern barge canal was in any case well away from the city center. Not long after, the salt industry disappeared as well. But Syracuse would continue to prosper even without salt: by 1918, its broadly based manufacturing sector produced everything from clocks and china to soda ash, shotguns, steam engines, men's shoes, and radiators.

At the canal's western terminus, Buffalo grew to be the greatest inland port in the United States, as Great Lakes shipping going east transferred its cargoes to the Erie Canal while canal traffic headed west was conveyed to ships waiting on Lake Erie. The huge volume of grain traffic from the west created the largest and most active grain-transfer port in the world there.

In 1842, Joseph Dart's invention of the grain elevator revolutionized the handling of wheat forever more. Before this, laborers down in the holds of lake cargo ships had to shovel the wheat into barrels, which Irish stevedores carried on their backs to warehouses. Dart's invention consisted of a steam engine moving a vertical belt to which buckets were attached. On the way down, the buckets went into the ship's hold and scooped up the grain; as the buckets reached the top of the device, they tilted over and dropped the grain into the warehouse awaiting its arrival.

Buffalo's access to both Lake Erie and the Erie Canal, and its strategic location almost equidistant from Chicago, New York, and Boston, led naturally to significant industrial development quite separate from the city's strategic role in handling grain. A quarter of a century later, it was a major rival of Pittsburgh in steel and after 1900 would also play a role in the development of the automobile industry.[44]

* * *

In all the towns and cities that boomed once the canal opened, a striking improvement in real wages and living standards accompanied the brisk growth in manufacturing and the accompanying improvement in labor productivity. For the first time, luxury articles were available to people in the working class. Clothes and even furniture were no longer homemade items but something to be bought in the stores, often mimicking the high fashion of the wealthy. One historian describes the pattern of social change as the "democracy of expectant capitalists."[45] Among a great mass of economic data, the most remarkable indicator of the improvement in living standards and aspirations is that an estimated 32,000 Irish women and free blacks were working as domestic servants in middle-class homes in New York City in 1855.[46]

If we were to end the story of the Erie Canal with the miracle it contributed to the unification and economic development of the United States, we would have a story of daring and determination, of enormous achievement and innovation, and of a cornucopia of riches developing across a unified nation. But all that would be only part of the story. The untold part is in many ways as heroic, as audacious, and as historic as what we have already noted. To recount one without the other would deprive each of its fullest meaning for modern times.

While Buffalo was clearly the western terminus of the canal, and the opening to the lands beyond, Albany at the eastern end has always been more of a transshipment facility than a terminus. The canal's true eastern limit was New York City, 150 miles to the south along the magnificent Hudson River. The story of the Erie Canal cannot come to an end without considering the profound importance of that connection. For it was not in the canal facilities at Albany that the Wedding of the Waters took place. Rather, the festivities culminated in the harbor of New York City, where Erie waters come down the Hudson to join the Atlantic Ocean.

CHAPTER 20

THE GRANARY OF THE WORLD

In April 1824, in the course of a speech about the future of the Erie Canal and New York City, De Witt Clinton produced another of his eerily accurate forecasts. After predicting that the most productive regions of America would use the canal for transporting goods abroad or for consumption at home, he went on to describe the glowing outlook for the city: "The city will, in the course of time, become the granary of the world, the emporium of commerce, the seat of manufactures . . . the concentrating point of vast disposable and accumulating capital, which will stimulate, enliven, extend and reward the exertions of human labor and ingenuity. . . . And before the revolution of a century, the whole island of Manhattan, covered with inhabitants and replenished with a dense population, will constitute one vast city."[1]

Yet Clinton's glowing phrases and startling foresight were incomplete, too parochial in their vision. Clinton senses what might develop, but a crucial chapter in the future of the Erie Canal and of New York City is missing. What is it precisely that will trigger the extraordinary level and quality of economic and financial activity he foresees for one little island of merely twenty-three square miles? That the canal would have an influence on the future of New York City was beyond question.

But the influence was reciprocal. New York City's vast port and diversified economy would have a dynamic impact on the towns and cities along the route of the Erie Canal, on the increasing thousands of people sailing west on the packets for a new life, and, indeed, on the canal itself.

Even then the story is unfinished. Clinton's remarkable foresight leaves out the most exciting part of the whole story. When he predicts that New York will become "the granary of the world," he provides us with no more than a hint of the transforming events that lie ahead. Rather, he leaves us to take his forecast on faith, without elaboration or support.

The enduring success of the canal was not just in the marvel of a waterway linking Lake Erie and Buffalo with New York, nor was it only in the impressive economic development it motivated between Buffalo and lands to the west of Lake Erie. "In the course of time," to borrow Clinton's expression, the Erie Canal would turn out to be the first great bridge between the inexhaustible supplies of grain from the midwestern United States and the inexhaustible demand for food from Europe—and Britain in particular.

But how could that happen? Europe had fed itself since the beginning of time. Why should Europeans be seeking sources across the seas now, after all those centuries of self-sufficiency? And why does it matter?

Put these questions to the pioneers of the Industrial Revolution in eighteenth-century Britain—most of whom were early investors in the canal network of the Midlands that had become the new industrial heartland of Britain. Ask James Watt and Matthew Boulton, who together made a reality of the steam engine, or Joseph Priestley, the discoverer of oxygen, or the pioneer of factory production Josiah Wedgwood, or the innovator-of-all-trades Erasmus Darwin. Ask the Duke of Bridgewater, who, as we noted earlier, built a canal to connect a coal mine directly to Manchester. Or, most important of all, consult the great classical economists Adam Smith and David Ricardo, whose work involved the most careful study of the relations between food supplies, wage rates, and a nation's prosperity.

As the driving force of nineteenth-century economic activity in Britain was transformed from agriculture to industry, and from animal power to steam power, management of food supplies for factory workers would develop into the critical variable of progress and the central focus of British politics and economic policy. Doing things the old way was no longer possible in an age of rapid-fire technological change. But in 1824 even De Witt Clinton could not foresee how vital the breadbasket of the midwest would be to the Industrial Revolution's tidal wave of change. Or that the Erie Canal would be the nexus of the network that made the whole thing possible.

* * *

When Clinton delivered his dazzling picture of the future of New York City, the case for New York as the nation's key city and port was by no means obvious. In 1664, after four English frigates sailed into the harbor and seized the community of New Amsterdam from the Dutch, the settlement was renamed in honor of the Duke of York. Although this event seems a far cry from the momentum of the Industrial Revolution in the 1800s, those four frigates freed New York from its limited role as an isolated Dutch trading enclave, transforming forever the entire character of the city's economy.

Even so, in terms of total tonnage of imports and exports during the colonial era, the port of New York stood only in fourth place after Philadelphia, Boston, and Charleston.[2] Then New York's development was stifled once the British took the port in the summer of 1776, when a substantial portion of the little community went up in flames. The Redcoats occupied New York for the duration of the war, right up to the signing of the Treaty of Paris in 1783; meanwhile, Boston, Philadelphia, Charleston, and Baltimore were free. Although New York's export business boomed after the Revolution and especially during the early years of the Napoleonic Wars after 1800, the export trade had prospered in the other cities as well.

One easy answer to the question of why New York City finally triumphed was its deep and expansive harbor, which was more attractive to shipping than the facilities at the other eastern seaboard ports. But that response is insufficient. The others also had fine harbors leading to the Atlantic Ocean. It was what lay *behind* New York City that ultimately made the difference to the history of the city: the Erie Canal, an unbroken link from the fertile lands at the heart of the United States to the Atlantic Ocean at New York. No other Atlantic port could even begin to match that.

The clincher for New York's leadership came in 1817, when the canal's construction began, which was solid evidence the canal was going to be a functioning and strategic waterway instead of just a politician's football. Although the canal would not reach full operational capability for another eight years, in the long run it brought greater and more durable prosperity to New York City than even the opening of the New York Stock Exchange, also in 1825.

After the completion of the canal and the Wedding of the Waters that year, there was no longer any doubt that New York was the port of choice in the United States and that New York would be the leading city of the nation and, in time, of the world. Money and finance centered in New York because of the huge volume of goods that entered and left the city every single day of the week. Its large and varied manufacturing enterprises made the New York area the prime source for a variety of industrial and manufactured products, ranging from machinery, steam engines, iron foundries, shipyards, newspaper printing, clothing, sugar refining, and boots and shoes to bread, crackers, and cabinet furniture.[3]

And goods moved in and out of New York because of its location between the Erie Canal and the sea. New York developed into such an important port that, right up to the Civil War, more than half of southern cotton production left for Liverpool and Le Havre from there. Most of the South's imports also entered the United States by way of New York.

Soon Clinton's words would echo over and over, as people increasingly referred to New York as "the great commercial emporium of America."[4]*

On July 4, 1828, Baltimore launched an attempt to overcome New York's advantages by reopening the route of George Washington's Patowmack Company with the Chesapeake & Ohio Canal (the project cost $14 million—double that of the Erie Canal). But lacking a gorge like Little Falls, the C & O still had to deal with crossing the mountains rather than cutting through them. Philadelphia constructed the Main Line, an elaborate and expensive combination of canal and railroad facilities to connect the city to Pittsburgh, but the project never functioned satisfactorily and was a financial disaster.

Bostonian entrepreneurs were smarter: they passed up canal construction altogether. Starting in 1833, they simply built the Western Railroad (known as the Boston & Albany after 1867) from Boston across the Berkshire Mountains to connect their spacious harbor to the Erie Canal's eastern terminus at Albany. Even that arrangement could not match what New York offered by way of water transportation—and, later on, railroad lines running parallel to the Erie Canal.

A few numbers can illustrate the difference in outcomes. In 1830, New York was already exporting four times as much as Philadelphia. By 1850, New York's exports had grown another 160 percent while Philadelphia's had stagnated. The same is true of exports out of Boston.[5]

* * *

The key to New York's spectacular growth has always been its access by waterway through that prehistoric gorge in Little Falls to the vast farmlands in the west. As historian Robert Greenough Albion has put it, New York took off as "the flour barrel began to replace the beaver skin

*The cognomen stuck. In his 1839 book, *Observations on the Financial Position and Credit of Such of the States of the North American Union as Have Contracted Public Debts*, Alexander Trotter, an English visitor, refers to New York as "the emporium" (p. 158).

as the port's most valuable offering to the world of commerce; and would remain so for more than a century."[6]

The full significance of replacing the beaver skin with the flour barrel would become visible only gradually, as New York became the dominant center of American trade and transportation, with the accompanying elaborate panoply of financial and commercial enterprise that followed. The real meaning of this development would not emerge until the middle of the nineteenth century, in a sequence of events that moved the world.

The story begins, not in the United States but in Britain, with the defeat of France and the return of peace in 1815 after many years of warfare.[7] The end of the Napoleonic Wars and of the intricate pattern of French and English blockades led to a return of freedom of the seas. This development was good news but there was bad news as well: the abrupt increase in supplies triggered a steep fall in commodity prices across Europe. Prices in Britain fell by a third between 1813 and 1820 and by another 25 percent by 1830. The British government took immediate action, enacting the Corn Law of 1815 (the word "corn" referred to all kinds of grain), essentially a set of protective tariffs designed to prevent cheap foreign food supplies from competing with British farm output, which the landed aristocracy had owned and controlled for centuries. As Adam Smith had expressed it so well in 1776 in *The Wealth of Nations*, "The landlords, like all other men, love to reap where they never sowed."

Legislation—Corn Laws—to discourage imports of food to Britain dated back to the late seventeenth century, but there was no need for a protective tariff as long as the British were able to export more corn and other grains than they imported. Indeed, productivity on British land had doubled and production had tripled from 1650 to 1800, as revolutionary a development as Britain's early leadership in the Industrial Revolution.[8] The last twenty-five years of the eighteenth century changed all that. The budding age of steam led to rapid growth of the industrial towns in the north and the Midlands, luring labor off the farms and into factory work. The result was an expansion in the demand for grains for human consumption, and England became a net importer of food for the first time in its history.

The enactment of the steep tariffs of the Corn Law of 1815 was the opening gun in what would develop into an extended battle to decide whether Britain would go on as it had for hundreds of years, as a largely agricultural economy in thrall to the landed aristocracy, or become instead the leader of the new more dynamic world of industry, manufacturing, and global trade. As Karl Marx would argue so effectively in the 1870s, food may come from farms, but its price is an integral part of the cost of industrial production because food keeps the urban working class alive and able to function in factories.

The nature of the conflict was clear. The aristocracy's standard of living depended on a high price for food. The competitive position and the profits of industrial entrepreneurs depended on a low price for food. The aristocracy was largely indifferent—even in many ways intolerant—to the social and political aspirations of the proletariat. Although industrial employers were reluctant to see workers gain political influence and economic bargaining power, they had less and less choice in the matter as the years progressed. Strikes and on occasion serious but aborted efforts at rebellion hit them where it hurt the most—in their pocketbooks.

When wartime exports of flour from the United States to Britain fell from 1.3 million barrels in 1812 to 620,000 barrels in 1816—and remained around that level for more than ten years, as a result of the Corn Law of 1815—the increase in the price of food provoked the British workers to demand higher wages.[9] The new capitalists, as employers, were not going to accept this blow without a fight. While they were prepared to resist labor demands, they recognized their more important objective was to repeal the Corn Laws that were the basic source of their troubles.

The field of battle was to be the House of Commons, where, from

1841 onward, the landlords' Tory party held a solid majority as well as the office of prime minister. The Tory majority did not mean that most British citizens favored a government dominated by the landed aristocracy. On the contrary, the large Tory majorities reflected an electoral system that discriminated against the rapidly growing number of the middle-class men who owned and managed the new factories, in addition to discriminating against their workers.* The theory behind this arrangement was as Daniel Defoe, the author of *Robinson Crusoe*, had put it in 1702: "I make no question but property of land is the best title to government in the world."[10]

Once launched, the attack on the Corn Laws turned out to be a stunning example of the law of unintended consequences. Forces for change broke out in tumult across a wide spectrum of Britain, lighting the fuse of discontent and frustration in a society going through revolutionary transformations in its power centers, in determining who had the right to vote, in the way people earned their living, and in the movement from the open country to the crowded slums of the cities. From the early 1830s to the late 1840s, Britain would go through a political and economic revolution every bit as profound and significant as the French Revolution at the end of the eighteenth century—although with far less bloodshed and much more gradual change.

Under the banner of Corn Law repeal, two Manchester manufacturers, Richard Cobden and John Bright, formed the Manchester Anti-Corn Law Association in 1836, which two years later combined with similar groups to compose the Anti-Corn Law League. Cobden and Bright's passionate assault on the Corn Laws had a much broader agenda than just the price of food. They perceived their battle as an integral part of the effort to transform Britain from a feudal society into a modern, aggressively competitive industrial nation bringing freedom and democ-

*The electoral system had also banned Catholics from voting. When parliamentary reform finally passed in 1838, Catholics—for the first time since Henry VIII—were given the right to vote and hold office.

racy for all. In a marathon sentence, Cobden described free trade as "breaking down . . . those barriers behind which nestle the feelings of pride, revenge, hatred, and jealousy which every now and then break their bonds and deluge whole countries with blood; those feelings which nourish the poison of war and conquest, which assert that without conquest we can have no trade, which foster that lust for conquest and dominion which sends forth your warrior chiefs to sanction devastation through other lands."[11]

Cobden led the attack against the Corn Laws with extraordinary zeal and tenacity. In October 1836, he declared, "It will only be done by a mighty effort of the irresistible masses. . . . *The Corn laws are only part of a system in which the Whig and Tory aristocracy have about an equal interest. The Colonies, the Army, Navy and Church, are, with the corn laws, merely accessories to our aristocratic government.*"[12] Three years later, he predicted, again in full-dress italics, "*We shall radicalize the country in the process of carrying the repeal of the Corn Laws. . . .* Our lecturers shall continue to haunt [the landlords] in their agricultural fastnesses, and our circulars shall proclaim their legislative robbery to the ends of the earth."[13] And this was a businessman, not Karl Marx!

The event Cobden worked so hard to bring about was a lot closer than the fifty years he was predicting. Bringing it about, however, involved nearly a decade of tumultuous struggles whose political battles and maneuvers made the conflict between De Witt Clinton and the Albany Regency twenty years earlier look like child's play.

The impetus that finally tipped the scale in favor of repeal came out of nowhere—disastrous crops throughout Europe in 1845 and on into 1846, causing a steep jump in food prices and widespread starvation among the poor. The most dramatic tragedy occurred in Ireland, where a blight of the potato crop left almost the entire land covered with black rot. The Irish peasants were hit from two sides. First, most of them were too poor to be able to afford anything except potatoes. In addition, the blight left them without the produce they needed to sell so they could pay the rent on their land to the absentee British Protestant landlords. Hundreds of thousands of Irish peasants were evicted. Those who survived starvation crowded into disease-infested cities or emigrated on ships on which disease and starvation also took a heavy toll. Even with British efforts to provide relief, at least 1 million Irish died from hunger and disease and another 2 million immigrated to America and Australia. By the end of the 1840s, the population of Ireland had shrunk from 8 million to 5 million.

Sir Robert Peel, the British prime minister, had always been sympathetic to the needs of the Irish, but he was the leader of the Tory party and as recently as 1841 had promised he would never repeal the Corn Laws. The catastrophic harvests of 1845 put the whole problem in a new light and forced him to change his mind on this overwhelmingly important issue. He was now convinced that repeal was essential, even if taking that position meant he would have to separate himself from his own party and join with the opposition Whigs. In a decisive speech to Parliament on January 22, 1846, Peel declared, "I am led to the conclusion that the main grounds of public policy on which protection has been defended are not tenable; at least I cannot maintain them."[14]

Although Peel and Cobden had been bitter enemies in the past, Peel now followed Cobden by exerting all his energies for repeal of the Corn Laws on the basis of social considerations much broader than just the immediate problem of ruinous harvests. He made this position clear in another speech before Parliament, in May, declaring that "all of you admit . . . we wish to elevate in the gradation of society that great class which gains its support by manual labor. The mere interests of the landlords [and] occupying tenants, important as they are, are subordinate to [that] great question."

Peel's choice of language supports the notion, held by many historians, that the Irish famine was a mere fig leaf for reform in the face of intense social pressures in Britain at that time. According to this view, Peel's motivation for drawing enough Tory votes to provide a majority for Corn Law repeal in 1846 was as much to forestall further social radicalism as it was specifically to benefit the industrial business class by opening British markets to imports of cheaper food from abroad.[15]

Even the Tories had to be aware that Europe was on the verge of revolution in 1846. Two years later, the fury of 1848 would erupt, with bloody revolutions in France and Germany, social commotion in Italy and Austria, and the publication of *The Communist Manifesto* by Karl Marx and Friedrich Engels. One evening in 1848, when Peel was sitting in the House of Commons and heard the news of the overthrow of the King of France, he said to his informant, "This comes of trying to govern the country through a narrow representation in Parliament, without regard to the wishes of those outside. It is what this party behind me wanted to do in the matter of the Corn Laws, and I would not do it."[16]

After bitter debates back and forth for a full six months, Peel finally swung the House of Commons in favor of repeal on May 15, 1846; the Lords followed a month later. Peel resigned shortly afterward, but his farewell speech to the House embellished the sentiments he had expressed a month earlier: "It may be that I shall leave a name sometimes remembered with good will in the abodes of those whose lot it is to . . . earn their daily bread by the sweat of their brow, when they shall recruit their exhausted strength with abundant and untaxed food, the sweeter because it is no longer leavened by a sense of injustice."

It was a great moment. From repeal of the Corn Laws until the depths of the Great Depression nearly a century later, Britain was the standard-bearer of free trade among the nations of the world. The result was a great leap forward in Britain's rate of economic growth. As the burden of tariffs was lifted, imports of corn tripled in just the first year after repeal, the price of food was lowered, and the reduction in the cost of food encouraged the migration away from agriculture and toward the cities and factories.[17] It was there that Britain's economic advantage was the greatest—in the coal-based industrial power of the Midlands and in the vast seagoing capacity that transported both export and import trade around the world. As employment at the factories soared, the share of agricultural workers in the British labor force shrank from 25 percent in 1846 to a little more than 10 percent by the end of the century.[18]

* * *

Now an interesting question arises. Except for the occasional and inevitable harvest failures, the British countryside had fed the nation for centuries without meaningful supplies of grain imports from abroad. Why were Peel and Cobden so confident they could take the risk of cutting the heart out of British agriculture without driving the price of food far higher than even the Corn Laws had done? Indeed, through all the spiteful and heated debates in the House of Commons, why did Peel's opponents never demand an answer from him on this matter?

The answer to these questions must be that both Peel and his opponents were confident the rest of the world could—and would—furnish a sufficient surplus of food production to offset whatever amounts would disappear from Britain's output as a result of free trade. But what made them so confident? The massive crop failures of 1845 and 1846 battered all of Continental Europe as well as the British Isles, but eastern Europe—primarily Russia and Poland—had been a significant source of British food imports. The best way to hedge risks is to diversify, but, clearly, relying on Europe's harvests to replace the British farmer was not diversification.

Although Britain also drew on agricultural output in the colonies, especially from Canada and even more distant places like Australia and India, the unspoken justification for taking the risk of deep cuts in home production must have been the burgeoning growth of agriculture in the new American west. But the American farmer, no matter how productive, would be no help on his own. What really mattered in these deliberations was the Erie Canal, the waterway that could bring the farmer's output over the hundreds of miles to the Atlantic ports at a minimal cost of transport.

No wonder the flow of American output to Britain surged after 1846 and never again dropped to the levels prevailing before the repeal of the Corn Laws. During the buildup to the War of 1812, both Jefferson and Madison figured that Europe needed America's food more than Americans needed luxuries from Europe, ignoring the damage the embargoes would inflict on American farmers, and the arrival of the war cast doubt over the whole diagnosis. But now, after the repeal of the Corn Laws—something neither Jefferson nor Madison could have anticipated—the swelling American exports of grain validated their viewpoint some thirty-five years after they had suggested it.

The data to support this hypothesis are impressive. During the ten years from 1836 to 1846, American merchandise exports to Britain averaged $48 million a year, with a high of $57 million in 1839. But in 1847—the first year after the repeal of the Corn Laws—Britain imported $87 million of merchandise from the United States. This was not just a temporary surge in response to the famines of the mid-1840s. From 1851 onward, there were only two years when American exports to Britain fell below $100 million, and even those were at $81 million and $92 million. It is also interesting to note that the prices on the New York Stock Exchange surged by 20 percent in 1846 and kept right on climbing for another four years.[19]

Although these exports include shipments of cotton from the South, much of the merchandise heading to Britain must have moved to the seacoast on the Erie Canal. From 1837 to 1845, the canal carried an average of 1.5 million tons of freight a year, with a high of 2 million in 1845. But in 1846, the year of repeal, the number moved up to 2.3 million tons and soared to just short of 3 million tons the following year. Canal tonnage continued to climb until it was running over 4 million tons in the 1850s. As we noted earlier, eastbound volume on the canal exceeded westbound volume for the first time in 1847—and the excess kept on growing. The importance of flour and grain in these shipments shows up in receipts of these commodities at Buffalo from the west from 1837 to 1860. Here, too, large upward jumps appear in 1846 and 1847 for both commodities; as we have seen, the tonnage reaching Buffalo from the west in the mid-1840s was ten times what it had been just ten years earlier.[20] By 1851, flour and grain shipments reaching Buffalo from the west were up to 18 million barrels and would reach 37 million by 1860.[21]*

Peel, Cobden, and their supporters had it right: repeal of the Corn Laws would not lead to a shortage of food in Britain.

* * *

In 1800, Gouverneur Morris had predicted that "one-tenth of the expense born by Britain in the last campaign would enable ships to sail from London through Hudson's river into Lake Erie." In his mind's eye, Morris fastened on what those British ships would be bringing to America. But the more interesting implications of his vision escaped him: the vast quantities of freight traveling on the return voyages, *from* Lake Erie through Hudson's river *to* London.

From George Washington onward, the dream motivating the construction of the Erie Canal was the vast inland navigation possible in the United States and the significance of that transportation system to the future of a country so large and so varied. Over and over, the exhortation came to cut through the mountains to bind a great nation into unity. But one of the great by-products of the Erie Canal was to knit the United States to Europe as well. In contemporary terms, globalization became the centerpiece of the story, as it is the centerpiece of so much economic activity and social change in our own time. New York City is still a primary center of the world economy a century later. Globalization is where the Wedding of the Waters renews its vows.

*The American farmer would literally save the day on two more highly dramatic occasions when Europe's crops failed later in the nineteenth century. See my *The Power of Gold: The History of an Obsession*, pp. 169 and 179.

EPILOGUE

We have traveled a long distance in this chronicle of the Erie Canal. In time, we have covered two and a half centuries, from Henry Hudson's abortive search for the Northwest Passage in 1609 to the eve of the Civil War in 1860. Over that long span of years, we have explored the vision that inspired the canal, the tenacious convictions that overwhelmed those who opposed it, the ingenuity that built and financed it, the appalling effort of human labor that created it, the boundless pride and enthusiasm of the young nation that welcomed it, and the profound economic change at home and abroad that was shaped by it.

The virtual space from the earliest vision of the canal to the full range of its consequences does not lend itself to measurement. Our survey has engaged us in a trip of almost unimaginable dimensions, from an old world clinging to the tradition of the ages to what Michael Chevalier so eloquently described as a new world of "circulation, motion, and boiling agitation." As the Founding Fathers created a new nation to "secure the blessings of Liberty to ourselves and our Posterity," the Erie Canal cut a waterway through the mountains to bind that nation into one and to make possible a new economic system in America that would meld forest, farm, and industry into a combination of extraordinary power.

One of the most striking features of the whole story is how much of it was part of the early visions. From the prophecies of George Washington to the predictions of De Witt Clinton, the great prizes to be gained by cutting a waterway through the mountains were national unity and economic power. Washington understood with remarkable clarity that unity was impossible without the economic side, without "commerce." Clinton was convinced that unity would be an inevitable consequence of economic achievement. Both understood that huge capital investments like these—no matter how financed and managed—would work only if they helped to make private markets function better. The final results were a shining tribute to both men's keen sense of the shape of the future. That a canal would bind the United States to Europe as well as joining eastern Americans to the west was a bonus Washington and Clinton may have sensed but that neither explicitly articulated.

It is ironic, from the vantage point of the early twenty-first century, that the two efforts to achieve these objectives—Washington's Patowmack Company and Clinton's Ditch—were so fundamentally different in structure, concept, and ultimate operational success. Although Clinton had the great advantage of geography—Virginia was not blessed with the gorge at Little Falls, creating a natural passage through the mountains—Washington's waterway was financed and managed as private enterprise while Clinton's was a public improvement from start to finish.

Suppose we had to bet today on which project would turn out to be the more successful effort, geography aside: a profit-seeking venture controlled by one of the great executives and administrators of all time, or a state-financed project managed by a committee of politicians. The choice seems to be an obvious one. Thomas Jefferson had reminded George Washington in early 1784, "Nature then has declared in favor of the Potomac, and through that channel offers to pour into our lap the whole commerce of the Western world. [Moreover] public undertakings are carelessly managed, and much money spent to little purpose."

Yet the privately owned and operated Patowmack Company ended up a financial failure and finished way behind schedule, while the committee of politicians who managed the construction of the Erie Canal would oversee their novel, complex, and gigantic project with high success, bringing it to completion on schedule, at a mind-boggling level of expenditure that came in close to original estimates, and without a single significant blunder or failure along the way. The long odds would have come out the big winner on that bet.

This striking contrast in final outcomes is all the more amazing when we recall that Washington was a trained surveyor with engineering experience. Perhaps he might have been able to carry it off if he had been able to enlist the resourcefulness, creativity, and single-minded devotion of Benjamin Wright, James Geddes, or Canvass White instead of the mercurial fraud James Rumsey. With no prior civil engineering experience, these men and their associates carried the Erie Canal up the Niagara escarpment at Lockport, maneuvered it onto a towering embankment to cross over Irondequoit Creek, spanned the Genesee River for it on an awesome aqueduct, and carved a route for it out of the solid rock between Little Falls and Schenectady—and all of those venturesome designs worked precisely as planned. We might also note that Washington ended up with a labor force of slaves, while the Erie Canal employed free men, including those who contributed anonymously to the technological achievements of construction, such as the machines to pull down enormous trees and then uproot their massive stumps.

When the construction of the canal was complete, and it was at long last time to celebrate the Wedding of the Waters, De Witt Clinton—the heroic protagonist of this story—spoke with more brevity and simplicity than with his usual flourishes. But he had the spirit of the occasion when he ended the ceremonies with these gentle words: "And may the God of Heavens and the Earth smile most propitiously on the work, and render it subservient to the best interests of the human race." His prayer was answered.

NOTES

Facts of publication for frequently used sources can be found in the Bibliography.

INTRODUCTION. "Does It Not Seem Like Magic?"

1. T. L. McKenny, quoted in Lionel Wyld, *Low Bridge!*, p. 27, citing Madeleine Waggoner, *The Long Haul West* (New York: G. P. Putnam's Sons, 1958), p. 113.
2. Wyld, *Low Bridge!*, p. 39, quoting from the diary of Lafayette's secretary, A. Levasseur, *Lafayette in America in 1824 and 1825.*
3. See Jeremy Atack and Peter Passell, *A New Economic View of American History from Colonial Times to 1940*, p. 166.
4. Douglass North. *The Economic Growth of the United States*, p. 253, Table E-Internet.
5. Blake McKelvey, *Rochester and the Erie Canal*, p. 18.
6. Don C. Sowers, *The Financial History of New York State*, p. 98.
7. David Hosack, *Memoir of De Witt Clinton*, appendix, note O. The passage is from a long letter to a friend in Hamburg, Germany.

CHAPTER 1. Smooth Sailing

1. Charles Hadfield, *World Canals*, p. 37.
2. John Phillips, *A General History of Inland Navigation*, p. 572.

3. For an extended description of flash locks, associated techniques, and later developments, see Hadfield, *World Canals*, pp. 30–37.
4. Ibid. pp. 16 and 417.
5. Most of the material for the Grand Canal comes from ibid., pp. 19–23.
6. Available at www.chinapage.com/canal.html.
7. The most interesting discussion I found on the subject of Leonardo's work in this area appears in an obscure and now defunct English journal featuring book reviews, *The Quarterly Review* 73 (1844), no. 146, especially pp. 293–97, where the author, Francis Egerton, cites works by writers named Frisi and Fumigalli. I found my copy of this journal at a small English bookseller as the result of a search on the Internet.
8. Rideau Waterway Co-ordinating Association. See www.rideaufriends .com/lockworks/lock-evolution.html.
9. Most of the material on the Canal du Midi comes from Odile de Roquette-Buisson and Christian Sarramon's *The Canal du Midi*, which is essentially a picture book with a large number of excellent photographs and contemporary diagrams of the canal's entire route, but also has first-rate commentary; the *Quarterly Review,* op. cit., p. 299; and Hadfield, *World Canals*, pp. 42–45.
10. Quoted in Hugh Malet, *The Canal Duke*, p. 31.
11. Roquette-Buisson and Sarramon, *Canal du Midi*, pp. 10–11, reproduces a magnificent map of 1726, with full details of the route and vivid drawings of the main structures along the way.
12. Hadfield, *World Canals*, pp. 42–43.
13. Most of the material on the Bridgewater Canal is from Malet, *Canal Duke*; Harold Bode, *James Brindley*; Hadfield, *World Canals*, pp. 57ff; and Jim Shead's magnificent site on the Internet, http://easy web.easynet.co.uk/jim.shead/index.htm or http://easyweb.easynet .co.uk/jim.shead/Bridgewater-Canal.html#BRDG, but the *Quarterly Review*, pp. 301–20, is truly delectable reading.
14. Malet, *Canal Duke*, p. 85.
15. Ibid., p. 88.
16. Elkanah Watson, *Men and Times of the Revolution . . .* , p. 164.
17. Ibid., p. 94.

CHAPTER 2. Hudson's Wrong Turn

1. William Miller, *The Geological History of New York State*, p. 21.
2. Milton Klein, *The Empire State*, p. 178.
3. Ibid., p. 146.
4. David Hosack, *Memoir of De Witt Clinton*, appendix, note N; all subsequent Colden quotations are taken from this source.
5. Ibid., note Q, quoting from Christopher Colles's report of his trip west in 1785.
6. For an extended description of Moore's efforts, see John Rutherford, *Facts and Observations in Relation to the Origin and Completion of the Erie Canal.*
7. Ibid., p. 54.
8. Rutherford, *Facts and Observations.*
9. Mary Riggs Diefendorf, *The Historic Mohawk.*
10. C. E. Bennett, *Many Mohawk Moons*, p. 320.
11. "Some Historical Context," available at www.nysm.nysed.gov/research_collections/research/history/neck/context.html.
12. Diefendorf, *Historic Mohawk.*
13. "Contemptible": Klein, *Empire State*, p. 259; Parkman: www.Britannica.com, article on Iroquois Confederation.
14. David Yarrow, "The Great Law of Peace: New World Roots of American Democracy," September 1987, available at www.kahonwes.com/iroquois/document.html.
15. I am grateful to Richard Sylla for this information.
16. www.nativetech.org.
17. Quoted in "Some Historical Context."
18. Ibid.

CHAPTER 3. Washington's Pivot

1. Cadwallader Colden, *Memoir at the Celebration. . . .*
2. De Witt Clinton, *Memorial . . . in Favour of a Canal Navigation.*

3. Quoted in Stacy Schiff, "Vive l'Histoire," *The New York Times*, February 6, 2003.
4. Quoted in James Flexner, *Steamboats Come True*, p. 384.
5. Quoted in ibid., p. 65.
6. David Hosack, *Memoir of De Witt Clinton*, appendix, note P.
7. Nathan Miller, *The Enterprise of a Free People*, p. 4, citing Washington's *Writings*, vol. 28, p. 127.
8. Ibid., p. 66.
9. http://memory.loc.gov/ammem/gmdhtml/gwmaps.html.
10. Rhoda Blumberg, *What's the Deal?*, p. 725.
11. Charles Hadfield, *World Canals*, p. 274.
12. Cynthia Owen Philip, *Robert Fulton*, p. 11.
13. Ibid., p. 67.
14. Ibid., p. 69. See also Ronald Shaw, *Erie Water West*, p. 12.
15. Quoted in World Regional Geography, "Technology, the Patowmack Canal, and National Unity," www.geog.okstate.edu.
16. John Marshall's recollections as contributed to Hosack, *Memoir*.
17. In 1791, Secretary of the Treasury Alexander Hamilton would define the metallic content of the dollar in terms of both gold and silver, setting up an implicit fixed rate of exchange between the dollar and the pound sterling.
18. Carter Goodrich, *Government Promotion of American Canals and Railroads*, p. 21.
19. Blumberg, *What's the Deal?* For more detail, see http://nps.gov/thst/mtver.htm.
20. http://nps.gov.thst.mtver.htm.
21. Quoted in Flexner, *Steamboats Come True*, p. 87.
22. Quoted in Blumberg, *What's the Deal?*, p. 744.
23. Quoted in Flexner, *Steamboats Come True*, p. 88.
24. Ibid.
25. Quoted in ibid., p. 98.
26. Blumberg, *What's the Deal?*, p. 730.

CHAPTER 4. Canal Maniacs

1. David Hosack, *Memoir of De Witt Clinton*, appendix, note Q.
2. Charles Merguerian, Hofstra University, lecture on history and geology of the New York City aqueduct system, 2000, available at www.duke labs.com/NYC%20Water%20Supply/NYCWaterSupply.htm.
3. This quotation and everything in the next three paragraphs is from Hosack, *Memoir*.
4. Ronald Shaw, *Erie Water West*, p. 13.
5. Quoted in Noble E. Whitford, *History of the Canal System of the State of New York*.
6. Quoted in Hosack, *Memoir*.
7. Elkanah Watson, *History of the Rise, Progress and Existing Condition . . .*, p. 175.
8. Quoted in Hosack, *Memoir*, appendix, note S.
9. Watson, *History of the Rise*, p. 164.
10. Ibid., pp. 243–46.
11. Ibid., p. 60.
12. Ibid., p. 7.
13. Ibid., p. 269.
14. Ibid., p. 271.
15. Ibid., p. 272. Italics in original.
16. Ibid., p. 274. Italics in original.
17. Ibid., pp. 15–16.
18. Ibid., p. 286.

CHAPTER 5. "A Canal to the Moon"

1. Elkanah Watson, *The Expedition*, p. 95.
2. From the national census of 1820. See Ronald Shaw, *Erie Water West*, p. 5.
3. See Watson, *Expedition*, pp. 19–50.
4. Ibid., pp. 57–58.

5. Ibid., p. 100.
6. Ibid., p. 19.
7. Ibid., p. 22.
8. Timothy Dwight, *Travels in New-England and New-York*, vol. 4, p. 124.
9. Nathan Miller, *The Enterprise of a Free People*, p. 23, citing Report of the Canal Commission, January 31, 1818.
10. Ibid., p. 26.
11. By far the most complete and rewarding early history of the Western Inland Lock Navigation Company is found in Philip Lord Jr., *The Navigators*, which also contains a wealth of well reproduced contemporary maps, accounts, and diagrams of the entire route from Albany to Oneida Lake.
12. Letter of March 14, 1792, from Philip Schuyler to Elkanah Watson, in Watson, *History of the Rise, Progress and Existing Condition . . .*, p. 318.
13. Miller, *Enterprise of a Free People*, p. 12.
14. The New York State Museum in Albany has a fascinating facsimile edition of the detailed survey in 1792 by the Western Company, led by Philip Schuyler, of the Mohawk River from Schenectady to Wood Creek.
15. Quoted in Evan Cornog, *The Birth of Empire*, p. 106.
16. Quoted in Shaw, *Erie Water West*, pp. 18–19.
17. Ibid., p. 19.
18. Ibid., p. 18, citing two histories of the area.
19. For a detailed, fascinating description of these boats and their many uses, see www.tencrucialdays.com/html/durham.htm.
20. Shaw, *Erie Water West*, pp. 17–18.
21. Ibid., p. 18. See also Miller, *Enterprise of a Free People*, p. 11, fn. 17, and p. 13, fn. 22.
22. See Carol Sheriff, *The Artificial River*, p. 54.
23. See Jeremy Atack and Peter Passell, *A New Economic View of American History from Colonial Times to 1940*, pp. 147–49, on transport costs for a variety of commodities and the impact of shifting from road to

waterway on the distance producers could afford to transport their merchandise.

24. See Barbara Ann Chernow, *Robert Morris, Land Speculator, 1790–1801* (New York: Arno, 1978).
25. Ibid. See also Cadwallader Colden, *Memoir at the Celebration. . . .*
26. Shaw, *Erie Water West*, p. 23.
27. David Hosack, *Memoir of De Witt Clinton*, appendix, note O.
28. See William Chazenof, *Joseph Ellicott and the Holland Land Company*, p. 81.
29. Quoted in Sheriff, *Artificial River*, p. 17.
30. See Shaw, *Erie Water West*, p. 24, quoting an article Hawley wrote in 1841 for the *Ontario Messenger*, which was printed at Canandaigua. Noble E. Whitford, *History of the Canal System of the State of New York*, reports that Hawley had the brainstorm in his supplier's office rather than at dinner with Geddes. Dorothie Bobbé, *De Witt Clinton*, pp. 156–58, discusses Hawley's achievement, and Clinton's admiration, but makes no mention of his having been a jailbird. She says he studied up on the European canals, journeyed to Lake Erie, "made due obeisance to God for forming Niagara," then went home and wrote his essays.
31. Hosack, *Memoir*.
32. Ibid.
33. Ibid.
34. Ibid.

CHAPTER 6. The Sublime Spectacle

1. U.S. Department of Commerce, *Historical Statistics of the United States: Colonial Times to 1970*, vol. 2, tables Y335–Y336, p. 1104.
2. Albert Gallatin, *Report of the Secretary of the Treasury on the Subject of Public Roads and Canals*, p. 3.
3. Ibid., p. 4.
4. Ibid., p. 6.

5. Ibid., pp. 7–8.
6. Ibid., p. 8.
7. Adam Smith, *Lectures on Jurisprudence.*
8. Quoted in Noble E. Whitford, *History of the Canal System of the State of New York*, citing J. A. L. Ringwalt, *Development of Transportation Systems in the United States* (Philadelphia: n.p., 1888), pp. 41–42.
9. Quoted in James Renwick, *Life of De Witt Clinton*, pp. 12–13.
10. William Bernstein, *The Birth of Plenty*, p. 257, citing Jan de Vries, *Economy of Europe in an Age of Crisis, 1600–1750* (Cambridge: Cambridge University Press, 1976), pp. 169–70.
11. Gallatin, *Report of the Secretary*, p. 111.
12. Ibid., p. 112.
13. Ibid., p. 114. Italics in original.
14. See Jeremy Atack and Peter Passell, *A New Economic View of American History from Colonial Times to 1940*, table 6.2, p. 149, for dazzling examples of how reducing the cost of transportation revolutionizes the quantitative and qualitative features of an economy.
15. Gallatin, *Report of the Secretary*, p. 122.
16. Ibid., pp. 122–23.
17. Department of Commerce, *Historical Statistics of the United States*, vol. 2, table A7, p. 8 (population); tables U2 and U9, p. 866 (exports and imports); table Q418, p. 750 (tonnage); table W99, p. 959 (patents).
18. The description of Forman's role is from his own words and also from Wright's report, both in David Hosack, *Memoir of De Witt Clinton*, appendix, note U.
19. Whitford, *History of the Canal System*, ch. 24.
20. See William Chazenof, *Joseph Ellicott and the Holland Land Company*, p. 22.
21. William Campbell, *The Life and Writings of De Witt Clinton*, p. 147.
22. Chazenof, *Joseph Ellicott*, p. 209, from Second Census of the United States.

23. Ibid., p. 160.
24. See ibid., pp. 160–61, citing personal papers of Joseph Ellicott.
25. Hosack, *Memoir*, appendix, note U.
26. Ibid.

CHAPTER 7. The Extravagant Proposal

1. See Arthur Redford, *The Economic History of England*, pp. 88–93, for a detailed discussion of these developments.
2. www.Historycentral.com/documents/Nation.html.
3. William Chazenof, *Joseph Ellicott and the Holland Land Company*, p. 115, citing Henry Adams, *History of the United States of America . . .*, who in turn cited John Lambert, *Travels Through Canada and the United States in the Years 1806, 1807, 1808*, vol. 2, pp. 64 and 54.
4. All quotes from Eddy here are from David Hosack, *Memoir of De Witt Clinton*, appendix, note W.
5. Ibid.
6. Noble E. Whitford, *History of the Canal System of the State of New York*, p. 12.
7. Dorothie Bobbé, *De Witt Clinton*, p. 145.
8. Evan Cornog, *The Birth of Empire*, p. 106, citing Hosack and Renwick.
9. Hosack, *Memoir*, appendix, note X.
10. See Gustavus Myers, *The History of Tammany Hall*, p. 39.
11. Census data and Chazenof, *Joseph Ellicott*, appendix I.
12. Clinton's "Address Before the New York Historical Society on the Iroquois or Six Nations," of December 6, 1811, can be found in William Campbell, *The Life and Writings of De Witt Clinton*; quotation at pp. 252 and 266.
13. Bobbé, *De Witt Clinton*, p. 96.
14. For an interesting commentary on the design for New York versus Washington, D.C., see Andro Linklater, *Measuring America*.

CHAPTER 8. The Expedition

1. Roger Haydon, ed., *Upstate Travels*, p. 17, citing Alexander Bell, *Men and Things in America*, p. 40.
2. Cynthia Owen Philip, *Robert Fulton*, p. 193.
3. See ibid., pp. 224–26, although the whole chapter is worth careful reading. See also James Flexner, *Steamboats Come True*, ch. 14, which was the source for the quotations in this paragraph and the preceding paragraph.
4. William Campbell, *The Life and Writings of De Witt Clinton*, p. 30.
5. Ibid., p. 31.
6. Andy Olenick and Richard Reisem, *Erie Canal Legacy*, p. 36.
7. For an entertaining description of stagecoaches, see Haydon, *Upstate Travels*, pp. 19–20.
8. Ibid., pp. 56–57 and 137–38.
9. Ibid., p. 73.
10. Ibid., pp. 70–71.
11. Ibid., p. 72.
12. Ibid., p. 43.
13. See William Chazenof, *Joseph Ellicott and the Holland Land Company*, p. 96.
14. Ibid., p. 53.
15. Ibid., p. 69.
16. Ibid., p. 136.
17. Ibid., pp. 140–41.
18. David Hosack, *Memoir of De Witt Clinton*, appendix, note O.
19. Ibid., citing a pamphlet by John Rutherford called *Facts and Observations in Relation to the Origin and Completion of the Erie Canal.*
20. Ibid.
21. Ibid.
22. See Elkanah Watson, *The Expedition*, p. 100.
23. De Witt Clinton, *The Canal Policy of the State of New York*, p. 24.

24. See http://bioguide.congress.gov/scripts/biodisplay.pl?index=P000446.
25. Clinton, *Canal Policy*, p. 24.
26. See Dorothie Bobbé, *De Witt Clinton*, pp. 210–11.
27. This and other quotations from the commission's report are found in Nathan Miller, *The Enterprise of a Free People*, p. 32.
28. Richard Brookhiser, *Gentleman Revolutionary*, p. 189, citing the *Journal of the Senate.*
29. Quoted in Ronald Shaw, *Erie Water West*, p. 43, citing Ellicott's *Letterbooks.*

CHAPTER 9. War and Peace

1. Jabez Hammond, *The History of the Political Parties in the State of New-York*, vol. 1, p. 302.
2. Nathan Miller, *The Enterprise of a Free People*, p. 38.
3. U.S. Department of Commerce, *Historical Statistics of the United States: Colonial Times to 1970*, vol. 2, tables Y336–Y337, p. 1104.
4. Ronald Shaw, *Erie Water West*, p. 47, without citation but from the 1812 report of the commission.
5. Chapter 1, "The Tradition," of Miller's *Enterprise of a Free People* has an extended and highly interesting overview of economic development in New York State and the central role played by the state government in the process.
6. I draw heavily here from Shaw, *Erie Water West*, pp. 48–49.
7. John Rutherford, *Facts and Observations in Relation to the Origin and Completion of the Erie Canal.*
8. Charles Glidden Haines, *Considerations on the Great Western Canal from the Hudson to Lake Erie*, pp. 26–27.
9. Shaw, *Erie Water West*, pp. 49–50.
10. Hammond, *Political Parties*, vol. 1, p. 289.
11. See http://famousamericans.net/danieldtompkins.
12. Richard Brookhiser, *Gentleman Revolutionary*, citing Henry Adams's biography of John Randolph, who was then Morris's brother-in-law.

13. Department of Commerce, *Historical Statistics of the United States*, vol. 2, pp. 904–5, #U317 and U324.
14. Michael Chevalier, *Society, Manners, and Politics in the United States*, p. 225.
15. Brookhiser, *Gentleman Revolutionary*, p. 195, citing Morris's diary and correspondence.
16. Ibid.
17. Ibid., p. 194.
18. For a lively and compressed recital of the War of 1812, especially in New York, see Walter McDougall, *Freedom Just Around the Corner*, pp. 413–21.
19. http://bioguide.congress.gov/scripts/biodisplay.pl?index=P000446.
20. Quoted in Garry Wills, *James Madison*, p. 116.
21. Ibid.
22. Department of Commerce, *Historical Statistics of the United States*, vol. 2, p. 760.
23. See Wills, *James Madison*, pp. 126–27.
24. Ibid., p. 97.
25. William Chazenof, *Joseph Ellicott and the Holland Land Company*, p. 122.
26. See ibid., ch. 17, for a quick overview of these events.
27. Ibid., p. 129.
28. See Wills, *James Madison*, ch. 9, "Frigates and a Fresh Start," pp. 107–15. Two extended Web sites also provide great and exciting detail: www.jmu.edu/madison/center/main_pages/madison_archives/life/war1812/overview.htm and www.geocities.com/Broadway/Alley/5443/supfrig.htm.
29. De Witt Clinton, *The Canal Policy of the State of New York*, pp. 26 and 28.
30. Quoted in Chazenof, *Joseph Ellicott*, p. 54.
31. David Hosack, *Memoir of De Witt Clinton*, appendix, note O.
32. Haines, *Considerations*, p. 22. Also, see Miller, *Enterprise of a Free People*, p. 41, fn. 2.

33. Adam Smith, *Lectures on Jurisprudence.*
34. Haines, *Considerations*, p. 50.
35. Ibid., p. 57.
36. Hosack, *Memoir.*
37. De Witt Clinton, *Memorial . . . in Favour of a Canal Navigation*, pp. 28–41.

CHAPTER 10. The Shower of Gold

1. All these quotations are from Nathan Miller, *The Enterprise of a Free People*, pp. 66–67.
2. Noble E. Whitford, *History of the Canal System of the State of New York*, introduction.
3. Quoted in Lionel Wyld, *Low Bridge!*, p. 161.
4. William Chazenof, *Joseph Ellicott and the Holland Land Company*, pp. 209–10.
5. Evan Cornog, *The Birth of Empire*, p. 159. This whole paragraph draws in part on his discussion at that point.
6. See Miller, *Enterprise of a Free People*, p. 46, citing David Hosack, *Memoir of De Witt Clinton*, and Luther Severance, *The Holland Land Company* (Buffalo: Buffalo Historical Papers, 1924), pp. 89 and 91.
7. De Witt Clinton, *The Canal Policy of the State of New York*, p. 26.
8. Ibid., p. 27.
9. See Miller, *Enterprise of a Free People*, p. 45, for a more extended discussion of this point.
10. Ronald Shaw, *Erie Water West*, p. 63, citing a letter dated April 24, 1816.
11. Miller, *Enterprise of a Free People*, p. 53, citing *Annals of Congress*, 14 Cong., 2 Sess., p. 854.
12. Clinton, *Canal Policy*, p. 43.
13. Miller, *Enterprise of a Free People*, pp. 9 and 10.
14. Clinton, *Canal Policy*, p. 45.
15. See Hosack, *Memoir*, appendix, note O.

16. For a detailed description of the negotiations, see George Tibbits, appendix, note EE, in ibid.
17. See Mark Kurlansky, *Salt*, pp. 245–47.
18. Hosack, *Memoir*, appendix, note EE.
19. Whitford, *History of the Canal System*, ch. 24, and Wyld, *Low Bridge!*, p. 14.
20. Two thousand miles on foot sounds like an awesome task indeed. The source for this information is Whitford, *History of the Canal System*.
21. Dorothie Bobbé, *De Witt Clinton*, pp. 206–7, quoting letter to Henry Post, August 1816.
22. Clinton, *Canal Policy*, p. 44.
23. Quoted in Miller, *Enterprise of a Free People*, p. 64.
24. Whitford, *History of the Canal System*.
25. See Miller, *Enterprise of a Free People*, p. 60, including fn. 4.
26. Quoted in Shaw, *Erie Water West*, p. 62.
27. Clinton, *Canal Policy*, pp. 43–44.
28. Letter dated February 18, 1817, quoted in Shaw, *Erie Water West*, p. 69.
29. Whitford, *History of the Canal System*, p. 121.
30. Hosack, *Memoir*, appendix, note CC.
31. See Cornog, *Birth of Empire*, p. 160.
32. Martin Van Buren, *Autobiography*, p. 55.
33. www.15development.com/wfredk/vanburenletter.htm. There is some evidence that this letter was not written by Van Buren and is a latter-day hoax. See Robert J. McCloskey, "Bogus Letter," *Washington Post*, October 7, 1983, p. A22.
34. For an extended and most interesting study of Kent, see Chief Judge Judith Kay's address at the Chicago Law School at www.bartleby.com/65/eq/equity.html.
35. www.law.umkc.edu/faculty/projects/ftrials/amistad/AMI_BTHO.HTM.
36. Ibid.
37. Wyld, *Low Bridge!*, p. 9.

CHAPTER 11. Digging the Ditch

1. Alexis de Tocqueville, *Journey to America*, p. 129.
2. Ibid., pp. 321–22.
3. Noble E. Whitford, *History of the Canal System of the State of New York*, reports that Benjamin Wright's son, who assisted in the survey of 1816, reported "that he could count upon the fingers of one hand" the spots of ground then cultivated along the route of the survey between Rome and the Seneca River.
4. William Chazenof, *Joseph Ellicott and the Holland Land Company*, p. 210, provides population data for western New York from the census of 1820.
5. De Witt Clinton, *The Canal Policy of the State of New York*, p. 44.
6. Ronald Shaw, *Erie Water West*, p. 91.
7. See Chazenof, *Joseph Ellicott*, p. 174.
8. Don C. Sowers, *The Financial History of New York State*, p. 30.
9. Shaw, *Erie Water West*, p. 92.
10. See ibid., p. 90.
11. Whitford, *History of the Canal System*.
12. Cadwallader Colden, *Memoir at the Celebration. . . .*
13. Clinton, *Canal Policy*, p. 46.
14. Charles Glidden Haines, *Considerations on the Great Western Canal from the Hudson to Lake Erie*, p. 7.
15. Ibid., p. 42.
16. Shaw, *Erie Water West*, p. 95–96.
17. Ibid.
18. Clinton, *Canal Policy*, p. 48.
19. Quoted in Chazenof, *Joseph Ellicott*, p. 164, citing a letter from Ellicott to Simeon De Witt.
20. Nathan Miller, *The Enterprise of a Free People*, p. 58, see fn. 53.
21. See Chazenof, *Joseph Ellicott*, p. 172 and fn. 25.
22. Paul Evans, *The Holland Land Company*, p. 289.

23. Elkanah Watson, *History of the Rise, Progress and Existing Condition . . .*, p. 79.
24. Ibid., pp. 80–81, from which the quotes in the following two paragraphs are also taken.

CHAPTER 12. Boom, Bust, Bonds

1. Samuel Rezneck, *Business Depressions and Financial Panics*, p. 62.
2. Murray Rothbard, *The Panic of 1819*, p. 14.
3. Garry Wills, *James Madison*, pp. 138–39, citing Henry Adams's *History of the United States of America. . . .*
4. Rothbard, *Panic of 1819*, p. 11.
5. Rezneck, *Business Depressions*, p. 23.
6. Bray Hammond, *Banks and Politics in America from the Revolution to the Civil War*, p. 227. For a full and authoritative history of the development of U.S. financial markets and the banking system from 1790 to 1840, see Richard Sylla, "U.S. Securities Markets and the Banking System, 1790–1840," *Review of the Federal Reserve Bank of St. Louis*, May–June 1998.
7. Rothbard, *Panic of 1819*, pp. 10 and 13.
8. John Kenneth Galbraith, *Money*, p. 74, citing *The Adams-Jefferson Letters*, edited by Lester Cappon, vol. 2 (Chapel Hill: University of North Carolina Press, 1988), p. 424.
9. Hammond, *Banks and Politics*, p. 237. Hammond provides rich detail on this whole matter.
10. Ibid., p. 253.
11. Rothbard, *Panic of 1819*, p. 13.
12. Quoted in ibid., p. 16.
13. Data in this paragraph is from ibid., pp. 17–18.
14. Hammond, *Banks and Politics*, p. 259.
15. Rezneck, *Business Depressions*, p. 54.
16. Hammond, *Banks and Politics*, p. 259.
17. Rezneck, *Business Depressions*, p. 55.

18. Ibid.
19. Ibid., p. 56.
20. Claudia Goldin and Hugh Rockoff, eds., *Strategic Factors in Nineteenth Century American Economic History*, p. 173.
21. Douglass North, *The Economic Growth of the United States*, pp. 182–83.
22. Rezneck, *Business Depressions*, p. 57, citing data from the state comptroller's office.
23. William Goetzmann, Roger Ibbotson, and Liang Peng, "A New Historical Database for the NYSE 1815–1925," appendix I.
24. U.S. Department of Commerce, *Historical Statistics of the United States: Colonial Times to 1970*, vol. II, series #U8, pp. 865–66.
25. Rezneck, *Business Depressions*, p. 61.
26. Ibid., p. 60.
27. Ibid., p. 65.
28. Quoted in Ron Chernow, *The House of Morgan*, p. 322.
29. Rezneck, *Business Depressions*, p. 64.
30. Rothbard, *Panic of 1819*, p. 21.
31. Nathan Miller, *The Enterprise of a Free People*, p. 97.
32. Department of Commerce, *Historical Statistics of the United States*, vol. 2, series 715–16, p. 163. See also Goldin and Rockoff, *Strategic Factors*, graphs on pp. 48–51.
33. Rothbard, *Panic of 1819*, p. 20.
34. Miller, *Enterprise of a Free People*, p. 98.
35. Milton Klein, *The Empire State*, p. 176.
36. Miller, *Enterprise of a Free People*, p. 97.
37. Ronald Shaw, *Erie Water West*, p. 121, citing *Rochester Gazette*, October 10, 1820.
38. Walter Werner and Steven Smith, *Wall Street*, pp. 180–81, provide a full listing of all bonds issued for the canal from June 1817 to July 1825—a total of forty-two separate issues.
39. Miller, *Enterprise of a Free People*, p. 89. I have drawn heavily on Miller's description of the Bank for Savings.

40. Ibid., p. 90.
41. De Witt Clinton, *The Canal Policy of the State of New York*, p. 48.
42. Charles Glidden Haines, *Considerations on the Great Western Canal from the Hudson to Lake Erie*, p. 38.
43. Miller, *Enterprise of a Free People*, pp. 95–97.
44. Don C. Sowers, *The Financial History of New York State*, appendix IV, p. 336.
45. Miller, *Enterprise of a Free People*, p. 101, furnishes ample detail on these developments in a long footnote.
46. Ibid., p. 102.
47. Sowers, *Financial History*, appendix IV, p. 336.
48. See Evan Cornog, *The Birth of Empire*, p. 161.
49. Clinton, *Canal Policy*, p. 50.
50. Ibid.

CHAPTER 13. Rude Invective

1. Jabez Hammond, *The History of the Political Parties in the State of New-York*, vol. 1, pp. 458–62, has an excellent appraisal of Clinton's egotistical manners.
2. See ibid., pp. 325 and 332.
3. www.multied.com/elections/1800.html.
4. Quoted in Evan Cornog, *The Birth of Empire*, p. 119.
5. Quoted in Stanley Fischer, "Globalization and Its Challenges," *American Economic Review*, May 2003, p. 1.
6. Gustavus Myers, *The History of Tammany Hall*, p. 1.
7. Ibid., p. 5.
8. Ibid., p. 10.
9. Henry Adams, *History of the United States of America . . .*, p. 143.
10. Dixon Ryan Fox, *The Decline of Aristocracy in the Politics of New York*, p. 57, citing W. A. Duer, *Reminiscences of an Old Yorker* (New York: Printed for W. L. Andrews, 1867), p. 24.
11. Adams, *History*, p. 143.

12. Myers, *Tammany Hall*, p. vii.
13. Ibid., p. 20.
14. Quoted in Dorothie Bobbé, *De Witt Clinton*, p. 84.
15. Hammond, *Political Parties*, vol. 1, p. 186.
16. Joanne Freeman, *Affairs of Honor*. I recommend this book for an extended and interesting discussion of these tendencies.
17. For an extended and exciting description of this duel, see Bobbé, *De Witt Clinton*, pp. 89–92.
18. Ibid., p. 223, probably quoting Adams's *History*.
19. Hammond, *Political Parties*, vol. 1, pp. 461–62.
20. Personal correspondence, cited in Ronald Shaw, *Erie Water West*, p. 105.
21. Quoted in Milton Klein, *The Empire State*, p. 300.
22. Shaw, *Erie Water West*, p. 108. The ally was Gideon Granger.
23. Ibid., p. 103, quoting Clinton's address to the legislature in January 1820.
24. Ibid., p. 497.
25. Ibid., p. 108, quoting a letter from Clinton to Young.
26. See Hammond, *Political Parties*, vol. 1, p. 454.
27. Quoted in Cornog, *Birth of Empire*, pp. 148–49.
28. Hammond, *Political Parties*, vol. 1, p. 517.
29. Shaw, *Erie Water West*, p. 113, quoting the issue of February 15, 1820.
30. Cornog, *Birth of Empire*, pp. 139–42, has a lively account of this election campaign.
31. Quoted in Shaw, *Erie Water West*, p. 115.
32. Cornog, *Birth of Empire*, p. 147, quoting a letter to Henry Post.
33. Noble E. Whitford, *History of the Canal System of the State of New York*, p. 32, citing *Assembly Journal*, 1820, p. 671.
34. Quoted in Shaw, *Erie Water West*, p. 117.
35. Hammond, *Political Parties*, vol. 2, p. 89.
36. Cited in Bobbé, *De Witt Clinton*, p. 246.
37. www.lcweb.loc.gov, document 605 of the Thomas Jefferson Papers.

38. Cornog, *Birth of Empire*, p. 143.
39. Ibid., p. 146.
40. Hammond, *Political Parties*, vol. 2, p. 101; also cited in Cornog, *Birth of Empire*, p. 146.

CHAPTER 14. Unwearied Zeal

1. Noble E. Whitford, *History of the Canal System of the State of New York*, ch. 24, citing George Geddes, "The Erie Canal," in *Publications of the Buffalo Historical Society*, pp. 291–93.
2. Ibid.
3. Ibid., citing Introduction to *Public Documents Relating to New York Canals*, p. xiii.
4. Ibid.
5. Ibid., p. 33.
6. Ronald Shaw, *Erie Water West*, p. 99, citing *Assembly Journal*, p. 671.
7. Anonymous, "Notes on a Tour Through the Western Part of the State of New York" (originally in the magazine *Ariel*, 1829–1830), in Warren Tryon, *A Mirror for Americans*, vol. 1, pp. 104–13.
8. Ibid., p. 112.
9. A superb photograph of the station appears in Debbie Stack and Ronald Marquisee, *Cruising America's Waterways: The Erie Canal*, p. 766.
10. Sibyl Tatum, quoted in Carol Sheriff, *The Artificial River*, p. 61.
11. Tryon, *Mirror for Americans*, vol. 1, p. 111.
12. Quoted in Dorothie Bobbé, *De Witt Clinton*, p. 245.
13. See Whitford, *History of the Canal System*, ch. 24, and Lionel Wyld, *Low Bridge!*, p. 49.
14. Tryon, *Mirror for Americans*, vol. 1, p. 106.
15. See Andy Olenick and Richard Reisem, *Erie Canal Legacy*, p. 24.
16. Ralph Andrist, *The Erie Canal*, p. 41.
17. From Freneau's 1822 poem "Oh, The Great Western Canal of the State of New York."
18. Blake McKelvey, *Rochester and the Erie Canal*, pp. 5–7, has a lively and extensive description of these developments.

19. Quoted in Wyld, *Low Bridge!*, p. 39.
20. John Howison, "A Tour from Rochester to Utica, 1820," in *Upstate Travels*, ed. Roger Haydon, p. 137.
21. See Wyld, *Low Bridge!*, p. 43.
22. See Francis Kimball, *New York—The Canal State*, p. 11.
23. William Chazenof, *Joseph Ellicott and the Holland Land Company*, p. 96, citing an account by Margaret Louise Plunkett called "The Upstate Cities and Villages," which he found in Alexander Flick, ed., *History of the State of New York* (New York: Columbia University Press, 1933–1937), vol. 8, p. 56.
24. McKelvey, *Rochester and the Erie Canal*, p. 7.
25. Patricia Anderson, *The Course of Empire*, p. 20, citing Nathan Parker Willis, *American Scenery* (London: George Virtue, 1840), p. 129.
26. Ibid., p. 36, citing Cole's *Diary*.
27. Quoted in Bobbé, *De Witt Clinton*, p. 253.
28. See Olenick and Reisem, *Erie Canal Legacy*, p. 16.
29. For a more extended description, see David Hosack, *Memoir of De Witt Clinton*, appendix, note CC by William Stone.
30. Cadwallader Colden, *Memoir at the Celebration....*
31. See Evan Cornog, *The Birth of Empire*, pp. 143–44.
32. See Shaw, *Erie Water West*, pp. 166–68, for an excellent description of the growing tension in Clinton's position.

CHAPTER 15. A Noble Work

1. Noble E. Whitford, *History of the Canal System of the State of New York*, p. 39.
2. Quoted in Carol Sheriff, *The Artificial River*, p. 31. The merchant's name was Ira Blossom.
3. Frances Trollope, *Domestic Manners of the Americans*, ch. 32.
4. I have drawn heavily here on the excellent description in Andy Olenick and Richard Reisem, *Erie Canal Legacy*, pp. 195–96.
5. See Ronald Shaw, *Erie Water West*, p. 133.
6. Sheriff, *Artificial River*, p. 35.

7. Roger Haydon, ed., *Upstate Travels*, pp. 211–12, citing Henry Tudor, *Narrative of a Tour in North America* (London: James and Duncan, 1834), vol. 1, pp. 230–34.
8. Lionel Wyld, *Low Bridge!*, p. 43, citing Frederick Gerstaecker, *Wild Sports in the Far West* (New York: John W. Lovell, 1881).
9. David McCullough, *The Path Between the Seas*, pp. 529–30.
10. Ibid., p. 498.
11. Ibid., p. 250.
12. Ibid., p. 481.
13. See www.hlc.wny.org/buffalo.jpg.
14. www.middlebass.org/lake_erie_steam_boats_1935.shtml.
15. William Chazenof, *Joseph Ellicott and the Holland Land Company*, p. 178, citing Peacock's report.
16. Quoted in Dorothie Bobbé, *De Witt Clinton*, p. 253.
17. See Whitford, *History of the Canal System*, p. 51.
18. This paragraph draws heavily on Shaw, *Erie Water West*, p. 160.

CHAPTER 16. The Pageant of Power

1. William Campbell, *The Life and Writings of De Witt Clinton*, p. 363.
2. Quoted in Dorothie Bobbé, *De Witt Clinton*, p. 254.
3. Ibid., p. 275.
4. Ibid., pp. 255–57.
5. Quoted in Evan Cornog, *The Birth of Empire*, p. 147.
6. Ibid.
7. Quoted in Ronald Shaw, *Erie Water West*, p. 166.
8. Dixon Ryan Fox, *The Decline of Aristocracy in the Politics of New York*, p. 283.
9. Quoted in Jabez Hammond, *The History of the Political Parties in the State of New-York*, vol. 2, p. 159. Italics in original.
10. Quoted in Fox, *Decline of Aristocracy*, p. 290, fn. 3.
11. http://en.wikipedia.org/upload/f/f9/ElectoralCollege1824-Large.png.

12. Martin Van Buren, *Autobiography*, p. 143.
13. David Hosack, *Memoir of De Witt Clinton.*
14. Van Buren, *Autobiography*, p. 143.
15. See Gustavus Myers, *The History of Tammany Hall*, p. 65, and Cornog, *Birth of Empire*, p. 151.
16. See Shaw, *Erie Water West*, p. 175, citing Thurlow Weed's *Autobiography.*
17. Shaw, ibid., p. 167, citing Weed in *Rochester Telegraph*, January 7, 1823.
18. Quoted in Fox, *Decline of Aristocracy*, p. 296.
19. Quoted in Myers, *Tammany Hall*, pp. 66–67.
20. Quoted in Cornog, *Birth of Empire*, p. 152.
21. Craig Hanyan and Mary L. Hanyan, *DeWitt Clinton and the Rise of the People's Men.*
22. Quoted in Fox, *Decline of Aristocracy*, p. 283, fn. 1.
23. Van Buren, *Autobiography*, pp. 143–45. Italics in original.
24. www.library.thinkquest.org.
25. Herbert Hoover, *The Memoirs of Herbert Hoover: The Great Depression*, p. 30.

CHAPTER 17. The Wedding of the Waters

1. Dorothie Bobbé, *De Witt Clinton*, p. 279; her source is not cited, but from David Hosack, *Memoir of De Witt Clinton.*
2. Ibid., p. 157.
3. Cadwallader Colden, *Memoir at the Celebration. . . .*
4. Quoted in Don C. Sowers, *The Financial History of New York State*, p. 63, citing S. H. Sweet, *History of Canals*, Assembly Documents, (1863), vol. 1.

CHAPTER 18. No Charge for Births

1. See Nathan Miller, *The Enterprise of a Free People*, p. 115, fn. 1.
2. I have purloined the expression "artificial river" from Gouveneur Morris's use of the term in 1803.

3. Nathaniel Hawthorne, "The Canal Boat," *New-England Magazine*, no. 9 (December 1835), pp. 398–409. This is the source for the next two quotations as well.
4. David Wilkie, "A Canal Journey in 1834," in Haydon, *Upstate Travels*, pp. 145–49.
5. Anonymous, "A Mirror for Americans," accessible at www.history.rochester.edu/canal/bib/1829.
6. John Howison, "A Tour from Rochester to Utica in 1820," in *Upstate Travels*, ed. Roger Haydon, p. 139.
7. Ibid., p. 140.
8. Anonymous, "Notes on a Tour Through the Western Part of the State of New York" (originally in the magazine *Ariel*, 1829–1830), in Tryon, *Mirror for Americans*, vol. 1, p. 114.
9. Robert G. Albion, *The Rise of New York Port*, pp. 87–90.
10. Tryon, *Mirror for Americans*, vol. 1, p. 113.
11. Frances Trollope, *Domestic Manners of the Americans*, ch. 32.
12. Ibid.
13. Hawthorne, "Canal Boat."
14. David Wilkie, "A Canal Journey in 1834," in Haydon, *Upstate Travels*, pp. 145–49.
15. Anonymous, "A Mirror for Americans."
16. Hawthorne, "Canal Boat."
17. See Lionel Wyld, *Low Bridge!*, p. 29.
18. Ibid., citing Patrick Shirreff, *Tour Through North America*, and *The Diary of Philip Hone*.
19. Ibid., citing *Diary of Philip Hone*.
20. Tryon, *Mirror for Americans*, vol. 1, p. 113. The traveler was Thomas Woodcock.
21. Wyld, *Low Bridge!*, p. 69, citing Myron Adams (Samuel Hopkins Adams's grandfather), *Grandfather Stories*.
22. Ibid., p. 18.
23. Data for the boat captain and team driver from Ronald Shaw, *Erie Water West*, p. 198, fn. 3.

24. Anonymous, "A Mirror for Americans." The misspelling of "berth" was not at all unusual in these commentaries.
25. See Wyld, *Low Bridge!*, p. 19, quoting from Clifton Johnson, *Highways and Byways of the Great Lakes* (New York: Macmillan, 1911), p. 30.
26. See Shaw, *Erie Water West*, p. 428, citing *Report of the Select Committee of the Assembly of 1846 upon the Investigation of Frauds . . . upon the Canals of the State of New York*, p. 348.
27. Michael Chevalier, *Society, Manners, and Politics in the United States*, p. 227.
28. Trollope, *Domestic Manners*.
29. Hawthorne, "Canal Boat."
30. Haydon, *Upstate Travels*, p. 196, citing Basil Hall, *Travels in America, in the Years 1827 and 1828*, 3rd ed. (Edinburgh: Cadell, 1829).
31. Carol Sheriff, *The Artificial River*, p. 98.
32. Whitney Cross, *The Burned-Over District*, pp. 85 and 86, provides interesting maps showing the decline in home-manufactured textiles in New York between 1825 and 1845.
33. See Sheriff, *Artificial River*, pp. 138–39.
34. Ibid., p. 142.
35. See ibid., p. 147, fn. 15, citing *Sailor's Magazine*.
36. See Shaw, *Erie Water West*, p. 221, fn. 8.
37. Herman Melville, *Moby-Dick*, ch. 54 ("The Town Ho's Story").
38. Sheriff, *Artificial River*, p. 143.
39. I borrow this expression from the explanation provided by Cross, *Burned-Over District*, p. 3.
40. See ibid., p. 80.
41. Sheriff, *Artificial River*, p. 121.
42. See Fareed Zakaria, *The Future of Freedom*, pp. 206–7, for an interesting analysis of these developments.
43. See Milton Klein, *The Empire State*, pp. 342–43.
44. See Cross, *Burned-Over District*, pp. 113–25, for an exhaustive account of this episode. See also James Morone, *Hellfire Nation: The Politics of Sin in American History* (New Haven: Yale University Press, 2003).

45. Elkanah Watson, *History of the Rise, Progress and Existing Condition . . .*, p. 22.

CHAPTER 19. The Prodigious Artery

1. Blake McKelvey, *Rochester and the Erie Canal*, p. 18.
2. See ibid.
3. Don C. Sowers, *The Financial History of New York State*, p. 98.
4. See Dorothie Bobbé, *De Witt Clinton*, p. 297.
5. See www.history.rochester.edu/canal/bib/nys1961/historyc.htm, and also Noble Whitford's *History of the Canal System of the State of New York*.
6. Cited by Jabez Hammond, *The History of the Political Parties in the State of New-York*, vol. 1, as a "rallying cry," p. 327.
7. Michael Chevalier, *Society, Manners, and Politics in the United States*, pp. 74 and 299.
8. Ibid., p. 130. Italics in original.
9. Ibid., pp. 282–83.
10. Ibid., p. 97.
11. Nathan Miller, *The Enterprise of a Free People*, pp. 198–99.
12. U.S. Department of Commerce, Bureau of the Census, *Historical Statistics of the United States: Colonial Times to 1970*, vol. 1, tables A-57 to A-72, p. 12.
13. Milton Klein, *The Empire State*, p. 289. For a further interesting contemporary analysis of the impact of the Erie Canal on the population, agriculture, and economic growth of New York State, see Alexander Trotter, *Observations on the Financial Position and Credit . . .*, pp. 84–86.
14. See Klein, *Empire State*, pp. 289–90.
15. Quoted in Ronald Shaw, *Erie Water West*, p. 277.
16. See http://xroads.virginia.edu/~HYPER/DETOC/TOUR/bufftxt.html.
17. Trotter, *Observations*, p. 85, and Sowers, *Financial History*, pp. 332–33.
18. Klein, *Empire State*, p. 315.

19. Douglass North, *The Economic Growth of the United States*, table L-IX, p. 257. Here, the west includes Illinois, Indiana, Iowa, Kansas, Kentucky, Michigan, Minnesota, Missouri, Nebraska, Ohio, Tennessee, Wisconsin, California, Nevada, and Oregon.
20. A stimulating account of this transformation (with many interesting citations to other works) appears in Algie Martin Simons, *Social Forces in American History*.
21. See Shaw, *Erie Water West*, pp. 264–65.
22. Ibid., p. 413, quoting Azariah C. Flagg papers.
23. Elkanah Watson, *History of the Rise, Progress and Existing Condition* . . . , pp. 15–16.
24. For an extended and often fascinating account of the canal as a force for economic development, see Miller, *Enterprise of a Free People*, chs. 7–12.
25. For more detail on the Illinois and Michigan Canal, see http://nps.gov/ilmi/.
26. Evan Cornog, *The Birth of Empire*, p. 162.
27. See Shaw, *Erie Water West*, p. 261.
28. See Jeremy Atack and Peter Passell, *A New Economic View of American History from Colonial Times to 1940*, p. 13.
29. North, *Economic Growth*, table E-IX, p. 253.
30. Francis Lieber quoted in Henry Steele Commager, *America in Perspective*, p. 33.
31. Quoted in North, *Economic Growth*, p. 173.
32. Kenneth Sokoloff, *Inventive Activity in Early Industrial America*, p. 14.
33. Ibid., p. 17.
34. Ibid., p. 10.
35. Quoted in Miller, *Enterprise of a Free People*, p. 138.
36. See Klein, *Empire State*, p. 314, where he quotes (but does not cite) the agricultural historian David Ellis.
37. Carter Goodrich, *Government Promotion of American Canals and Railroad*, p. 350, n. 30.
38. Dixon Ryan Fox, *The Decline of Aristocracy in the Politics of New York*, ch. 10.

39. Klein, *Empire State*, p. 318, citing federal and state census data.
40. See Alfred D. Chandler Jr., *The Visible Hand*, pp. 14–26.
41. Quoted in Blake McKelvey, *A Panoramic History of Rochester and Monroe County, New York*, p. 54.
42. Robert G. Albion, *The Rise of New York Port*, p. 89.
43. Quoted in Mark Kurlansky, *Salt*, p. 248.
44. In addition to using a variety of material found on the Internet, I have drawn on Andy Olenick and Richard Reisem, *Erie Canal Legacy*, in these sketches of Rochester, Syracuse, and Buffalo.
45. Simons, *Social Forces*, p. 210.
46. Klein, *Empire State*, p. 326.

CHAPTER 20. The Granary of the World

1. From an address delivered to a meeting of citizens in Albany on April 28, 1824, and reprinted as an appendix in David Hosack, *Memoir of De Witt Clinton*.
2. Robert G. Albion, *The Rise of New York Port*, p. 5.
3. Douglass North, *The Economic Growth of the United States*, pp. 168–76, and in particular his heavy emphasis on America's accomplishments in free education and the nourishment of human capital. See also Milton Klein, *The Empire State*, ch. 17.
4. The expression appears in Albion, *Rise of New York Port*, p. 15, but it crops up elsewhere in the literature as well.
5. Algie Martin Simons, *Social Forces in American History*, citing *Hunt's Merchant Magazine*, August 1868, p. 113.
6. Albion, *Rise of New York Port*, pp. 2 and 9–10.
7. I have drawn much important information on this subject from the work of Cheryl Schonhardt-Bailey of the London School of Economics, a distinguished expert in this field—in particular, from "Free Trade's Last Hurdle: Repeal of the Corn Laws in the House of Lords," a paper presented at the 2003 American Political Science Association Annual Meeting and scheduled to appear in her forthcoming book *Interest, Ideas and Institutions: Repeal of the Corn Laws Re-Told*. See

also her "Conservatives Who Sounded Like Trustees but Voted Like Delegates" (2002), for which the full text is available at http://personal.lse.ac.uk/schonhar/paper%202.pdf.

8. For a provocative analysis of the British agricultural revolution as compared with the "new economy" of our own time, see J. Bradford Delong, "A Historical Perspective on the New Economy."
9. Albion, *Rise of New York Port*, p. 92.
10. Quoted in Schonhardt-Bailey, "Free Trade's Last Hurdle."
11. Quoted in Michael Howard, *War and the Liberal Conscience*, pp. 42–43, citing Cobden's *Speeches on Questions of Public Policy* (London: n.p., 1870), vol. 1, p. 79.
12. Quoted in Wendy Hinde, *Richard Cobden*, p. 61. Italics in original.
13. Ibid., p. 74.
14. This and subsequent quotations from Peel may be found at http://dspace.dial.pipex.com/town/terrace/adw03/polspeech/speetop.htm, a site rich in material about Peel.
15. See Schonhardt-Bailey, "Free Trade's Last Hurdle"; I am grateful for her suggestion that the price of corn may not have been the primary motivation for Corn Law repeal.
16. http://dspace.dial.pipex.com/town/terrace/adw03/polspeech/speetop.htm.
17. See B. R. Mitchell, *European Historical Statistics*, various tables.
18. William Bernstein, *The Birth of Plenty*, p. 205, citing Angus Maddison, *The World Economy: A Millennial Perspective*, pp. 241 and 261.
19. William Goetzmann, Roger Ibbotson, and Liang Peng, "A New Historical Database for the NYSE 1815–1925," appendix I.
20. North, *Economic Growth*, table E-IX, p. 253.
21. Percy Bidwell and John Falconer, *History of Agriculture in the Northern United States, 1620–1860*, p. 310, citing the Buffalo Board of Trade.

BIBLIOGRAPHY

I have made extensive use of the Internet in carrying out the research for this book and, where appropriate, have identified URLs. Much of it was original material by witnesses to the story I have told. In those instances in which my sole source was the Internet rather than hard copy, I have been unable to provide the usual page references in my endnotes.

Two major Internet sources were especially important and useful.

Dr. David Hosack, a world-renowned physician and horticulturalist, was among De Witt Clinton's closest friends as well as his personal doctor. His *Memoir of De Witt Clinton*—which appeared only a little more than a year after Clinton's death in 1828—includes essays from every individual who had played a role of any importance, direct or indirect, in the development of the canal, reciting their views of the events and of the actions of the leading characters in the drama. The result is a rare historical narrative, a lively and colorful description of the construction of the canal as well as the political battles it ignited. I have made generous use of this extraordinary contemporary evidence.

The other work is the history of the canal by Noble E. Whitford, who was resident engineer at the New York State Engineer's Department when his *History of the Canal System of the State of New York* . . . was published in 1905. This work was the first extended history of the Erie Canal and is outstanding for its coverage, detail, and lucidity. All subsequent studies of any quality have drawn heavily on Whitford's effort.

BOOKS AND PAMPHLETS

Adams, Henry. *History of the United States of America During the Administrations of Jefferson and Madison.* 1921; reprint, Chicago: University of Chicago Press, 1967.

———. *The Life of Albert Gallatin.* London: J. B. Lippincott, 1880.

Albion, Robert G. *The Rise of New York Port: 1815–1860.* New York: Charles Scribner's Sons, 1939.

Anderson, Patricia. *The Course of Empire: The Erie Canal and the New York Landscape.* Rochester: Memorial Art Gallery of the University of Rochester, 1984.

Andrist, Ralph. *The Erie Canal.* New York: American Heritage Publishing, 1964.

Atack, Jeremy, and Peter Passell. *A New Economic View of American History from Colonial Times to 1940*, 2nd ed. New York: W. W. Norton, 1994.

Bell, Andrew ("A. Thomason"). *Men and Things in America; Being the Experience of a Year's Residence in the United States.* London: W. Smith, 1838.

Bennett, C. E. *Many Mohawk Moons.* Albany: self-published, 1938.

Bernstein, Peter L. *The Power of Gold: The History of an Obsession.* New York: Wiley, 2000.

Bernstein, William. *The Birth of Plenty: How the Prosperity of the Modern World Was Created.* New York: McGraw-Hill, 2004.

Bidwell, Percy, and John Falconer. *History of Agriculture in the Northern United States, 1620–1860.* New York: Peter Smith, 1941.

Blumberg, Rhoda. *What's the Deal? Jefferson, Napoleon, and the Louisiana Purchase.* Washington, D.C.: National Geographic Society, 1988.

Bobbé, Dorothie. *De Witt Clinton.* New York: Minton, Balch, 1933.

Bode, Harold. *James Brindley: An Illustrated Life of James Brindley, 1716–1772.* Princes Risborough, Eng.: Shire Publications, 1973.

Boucher, Cyril. *James Brindley: Engineer, 1716–1772*. Norwich, Eng.: Goose & Son, 1968.

Brookhiser, Richard. *Gentleman Revolutionary: Gouverneur Morris—The Rake Who Wrote the Constitution*. New York: Free Press, 2003.

Campbell, William. *The Life and Writings of De Witt Clinton*. New York: Baker and Scribner, 1849.

Chalmers, Harvey, II. *How the Irish Built the Erie*. New York: Bookman Associates, 1964.

Chandler, Alfred D., Jr. *The Visible Hand: The Managerial Revolution in American Business*. Cambridge, Mass.: Belknap Press, 1977.

Chazenof, William. *Joseph Ellicott and the Holland Land Company: The Opening of Western New York*. Syracuse: Syracuse University Press, 1970.

Chernow, Ron. *The House of Morgan: An American Banking Dynasty and the Rise of Modern Finance*. New York: Simon & Schuster, 1990.

Chevalier, Michael. *Society, Manners, and Politics in the United States: Letters on North America*. 1835; reprint, New York: Anchor, 1961.

Clark, Kenneth. *Leonardo da Vinci*. New York: Penguin, 1993.

Clinton, De Witt. *Memorial, of the Citizens of New-York, in Favour of a Canal Navigation Between the Great Western Lakes and the Tide-waters of the Hudson*. Albany: Bosford, 1816. Facsimile reprinted by UMI Books on Demand.

——— ("Tacitus"). *The Canal Policy of the State of New York*. Albany: Bosford, 1821. Facsimile reprinted by UMI Books on Demand.

Colden, Cadwallader. *Memoir at the Celebration for the Completion of the New York Canals*. New York: W. A. Davis, 1825. Also available at www.history.rochester.edu/canal/bib/colden/Memoir.html.

Commager, Henry Steele, ed. *America in Perspective: The United States Viewed Through Foreign Eyes in Thirty-five Essays*. New York: New American Library, 1947.

Cornog, Evan. *The Birth of Empire: De Witt Clinton and the American Experience, 1769–1828*. New York: Oxford University Press, 1998.

Court, W. H. B. *A Concise Economic History of Britain from 1750 to Recent Times*. London: Cambridge University Press, 1964.

Cronon, William. *Nature's Metropolis: Chicago and the Great West.* New York: W. W. Norton, 1991.

Cross, Whitney. *The Burned-Over District: The Social and Intellectual History of Enthusiastic Religion in Western New York, 1800–1850.* Ithaca: Cornell University Press, 1950.

Darby, William. *A Tour from the City of New York to Detroit.* 1819; reprint, Chicago: Quadrangle, 1962.

Diefendorf, Mary Riggs. *The Historic Mohawk.* New York: Putnam, 1910.

Drago, Harry Sinclair. *Canal Days in America: The History and Romance of Old Towpaths and Waterways.* New York: Clarkson N. Potter, 1972.

Dwight, Timothy. *Travels in New-England and New-York.* 4 vols. New Haven: S. Converse, 1822.

Evans, Paul. *The Holland Land Company.* Buffalo: Buffalo Historical Society, 1924.

Finch, Roy. *The Story of New York State Canals: Historical and Commercial Information.* Albany: State of New York, State Engineer and Surveyor, n.d.

Flexner, James. *Steamboats Come True.* New York: Viking, 1944.

Fox, Dixon Ryan. *The Decline of Aristocracy in the Politics of New York: 1801–1840.* 1919; reprint, New York: Harper Torchbooks, 1965.

Freeman, Joanne. *Affairs of Honor: National Politics in the New Republic.* New Haven: Yale University Press, 2001.

Galbraith, John Kenneth. *Money: Whence It Came, Where It Went.* New York: Houghton Mifflin, 1975.

Gallatin, Albert. *Report of the Secretary of the Treasury on the Subject of Public Roads and Canals.* 1808; reprint, New York: Augustus M. Kelley, 1968.

Goldin, Claudia, and Hugh Rockoff, eds. *Strategic Factors in Nineteenth Century American Economic History.* Chicago: University of Chicago Press, 1992.

Goodrich, Carter. *Government Promotion of American Canals and Railroads, 1800–1890.* New York: Columbia University Press, 1960.

Gronowicz, Anthony. *Race and Class Politics in New York City Before the Civil War.* Boston: Northeastern University Press, 1998.

Hadfield, Charles. *World Canals: Inland Navigation, Past and Present.* New York: Facts-on-File, 1986.

Haines, Charles Glidden. *Considerations on the Great Western Canal from the Hudson to Lake Erie.* Brooklyn: Spooner & Worthington, 1818. Facsimile reprinted by UMI Books on Demand.

Hammond, Bray. *Banks and Politics in America from the Revolution to the Civil War.* Princeton, N.J.: Princeton University Press, 1957.

Hammond, Jabez. *The History of the Political Parties in the State of New-York.* 2 vols. Syracuse: Hall, Mills & Co., 1852.

Hanyan, Craig, and Mary L. Hanyan. *DeWitt Clinton and the Rise of the People's Men.* Montreal: McGill-Queen's University Press, 1996.

Haydon, Roger, ed. *Upstate Travels: British Views of Nineteenth-Century New York.* Syracuse: Syracuse University Press, 1982.

Hilts, Len. *Timmy O'Dowd and the Big Ditch: A Story of the Glory Days on the Old Erie Canal.* New York: Gulliver Books, 1988.

Hinde, Wendy. *Richard Cobden: A Victorian Outsider.* New Haven: Yale University Press, 1987.

Hine, Robert, and Edwin Bingham, eds. *The American Frontier: Readings and Documents.* Boston: Little, Brown, 1972.

Hoover, Herbert. *The Memoirs of Herbert Hoover: The Great Depression, 1929–41.* New York: Macmillan, 1952.

Howard, Michael. *War and the Liberal Conscience.* New Brunswick, N.J.: Rutgers University Press, 1986.

Hubbard, J. T. W. *For Each, the Strength of All: A History of Banking in the State of New York.* New York: New York University Press, 1995.

Kennedy, Roger. *Burr, Hamilton, and Jefferson: A Study in Character.* New York: Oxford University Press, 2000.

Kimball, Francis. *New York—The Canal State: The Story of America's Great Water Route from the Lakes to the Sea, Builder of East and West.* Albany: Argus Press, 1937.

Klein, Milton. *The Empire State: A History of New York.* Ithaca: Cornell University Press, 2001.

Kurlansky, Mark. *Salt: A World History.* New York: Penguin, 2002.

Lambert, John. *Travels Through Canada and the United States in the Years 1806, 1807, 1808*. London: Baldwin, Craddock, and Joy, 1810.

Linklater, Andro. *Measuring America*. New York: Walker, 2002.

Lombard, Alexandre. *The Present Financial Situation of New York State*, 1848. Geneva: Lombard Odier & Cie., n.d.

Lord, Philip, Jr. *The Navigators: A Journal of the Passage of the Inland Waterways of New York*, 1793. Albany: New York State Museum, 2003.

Maddison, Angus. *The World Economy: A Millenial Perspective*. Paris: Development Centre of the Organisation for Economic Co-operation and Development, 2001.

McBain, Howard. *De Witt Clinton and the Origin of the Spoils System in New York*. New York: AMS Press, 1967.

McCullough, David. *The Path Between the Seas: The Creation of the Panama Canal, 1870–1914*. New York: Simon & Schuster, 1977.

McDougall, Walter. *Freedom Just Around the Corner*. New York: HarperCollins, 2004.

McKelvey, Blake. *Old and New Landmarks and Historic Houses*. Rochester: Rochester Public Library, 1950.

———. *A Panoramic History of Rochester and Monroe County, New York*. Woodland Hills, Calif.: Windsor Publications, 1979.

———. *Rochester and the Erie Canal*. Rochester: Rochester Public Library, 1949.

———. *Rochester: The Water-Power City, 1812–1854*. Cambridge, Eng.: Cambridge University Press, 1945.

Malet, Hugh. *The Canal Duke: A Biography of Francis, Third Duke of Bridgewater*. London: David & Charles, 1961.

Melville, Herman. *Moby-Dick*. 1851; reprint, New York: Signet, 1961.

Mereness, Newton, ed. *Travels in the American Colonies*. New York: Antiquarian Press, 1961.

Miller, Nathan. *The Enterprise of a Free People: Aspects of Economic Development in New York State During the Canal Period, 1792–1838*. Ithaca: Cornell University Press, 1962.

Miller, William. *The Geological History of New York State*. Albany: New York State Museum, 1924 (Museum Bulletin 168).

Mitchell, B. R. *European Historical Statistics, 1750–1970.* New York: Columbia University Press, 1976.

Mokyr, Joel. *The Gifts of Athena: Historical Origins of the Knowledge Economy.* Princeton, N.J.: Princeton University Press, 2002.

Morton, Desmond. *A Short History of Canada,* 5th ed. Toronto: McClelland & Stewart, 2001.

Murphy, Dan. *The Erie Canal: The Ditch That Opened a Nation.* Buffalo: Western New York Wares, 2001.

Myers, Gustavus. *The History of Tammany Hall,* 2nd ed. New York: Boni & Liveright, 1917.

Nevins, Allan, ed. *America Through British Eyes.* New York: Oxford University Press, 1948.

Nordhaus, William D. *The Economic Consequences of a War in Iraq.* Cambridge, Mass.: National Bureau of Economic Research, 2002 (Working Paper #9361).

North, Douglass. *The Economic Growth of the United States: 1790–1860.* New York: W. W. Norton, 1966.

Olenick, Andy, and Richard Reisem. *Erie Canal Legacy: Architectural Treasures of the Empire State.* Rochester: The Landmark Society of Western New York, 2000.

Peña, Elizabeth. "Making 'Money' the Old-Fashioned Way: Eighteenth Century Wampum Production in Albany." In *People, Places and Material Things: Historical Archeology of Albany, New York,* edited by Charles L. Fisher. Albany: New York State Museum, 2003 (Museum Bulletin 499).

Philip, Cynthia Owen. *Robert Fulton.* New York: Franklin Watts, 1985.

Phillips, John. *A General History of Inland Navigation, Foreign and Domestic; Containing a Complete Account of the Canals, Already Executed in England with Considerations on Those Proposals.* 1805; reprint, New York, Augustus M. Kelley, 1970.

Population of New York State by County 1790 to 1990. Albany: New York State Department of Economic Development, 2000.

Powys, Llewelyn. *Henry Hudson.* New York: Harper & Brothers, 1928.

Read, Donald. *Cobden and Bright: A Victorian Political Partnership.* London: Edward Arnold, 1967.

Redford, Arthur. *The Economic History of England, 1760–1860*. London: Longmans Green, 1960.

Renwick, James. *Life of De Witt Clinton*. New York: Harper & Brothers, 1840.

Rezneck, Samuel. *Business Depressions and Financial Panics: Essays in American Business and Economic History*. New York: Greenwood Publishing, 1968.

Roquette-Buisson, Odile de, and Christian Sarramon. *The Canal du Midi*. New York: Thames and Hudson, 1981.

Shaw, Ronald. *Erie Water West: A History of the Erie Canal, 1792–1854*. Lexington: University of Kentucky Press, 1966.

Sheriff, Carol. *The Artificial River: The Erie Canal and the Paradox of Progress, 1817–1862*. New York: Hill & Wang, 1996.

Simons, Algie Martin. *Social Forces in American History*. 1911; reprint, Lawrence, Kans.: Carrie Books, 2003. Also available at www.ku.edu/carrie/texts/carrie_books/simons.

Smith, Adam. *Lectures on Jurisprudence*. Vol. 5 of *The Glasgow Edition of the Works and Correspondence of Adam Smith*. Edited by R. L. Meeks, D. D. Raphael, and P. G. Stein. Indianapolis: Liberty Fund, 1982.

Sokoloff, Kenneth. *Inventive Activity in Early Industrial America: Evidence from Patent Records, 1790–1846*. Cambridge, Mass.: National Bureau of Economic Research, 1988 (Working Paper #2707).

Sowers, Don C. *The Financial History of New York State*. 1914; reprint, New York: AMS Press, 1969.

Stack, Debbie, and Captain Ronald Marquisee. *Cruising America's Waterways: The Erie Canal*. Manlius, N.Y.: Media Artists, 2001.

Thomson, David. *The Pelican History of England: England in the Nineteenth Century*. London: Penguin Books, 1950.

Thorp, Willard. *Business Annals*. New York: Garland Publishing, 1983.

Tocqueville, Alexis de. *Journey to America*. 1832; reprint, New Haven: Yale University Press, 1960.

Trollope, Frances. *Domestic Manners of the Americans*. 1832; reprint, New York: Penguin, 1997. Chapter 32 is also available at www.history.rochester.edu/canal/bib.

Trotter, Alexander. *Observations on the Financial Position and Credit of Such of the States of the North American Union as Have Contracted Public Debts.* 1839; reprint, New York: Augustus M. Kelley, 1968.

Tryon, Warren. *A Mirror for Americans: Life and Manners in the United States, 1799–1870.* Vol. 1. Chicago: University of Chicago Press, 1952.

Turner, Frederick Jackson. *The Frontier in American History.* New York: Henry Holt, 1920.

Uglow, Jenny. *The Lunar Men: Five Friends Whose Curiosity Changed the World.* New York: Farrar Straus and Giroux, 2002.

U.S. Department of Commerce, Bureau of the Census. *Historical Statistics of the United States: Colonial Times to 1970.* 2 vols. Washington, D.C.: Government Printing Office, 1975.

Van Buren, Martin. *Autobiography.* Edited by John C. Fitzpatrick. Washington, D.C.: Government Printing Office, 1920.

Watson, Elkanah. *History of the Rise, Progress and Existing Condition of the Western Canals in the State of New York, from September 1788 to the Completion of the Middle Section of the Grand Canal in 1818.* Albany: D. Steele, 1820. Facsimile reprinted by UMI Books on Demand.

———. *Men and Times of the Revolution or Memoirs of Elkanah Watson, Including Journal of Travels in Europe and America from 1777 to 1842, Edited by His Son Winslow Watson.* New York: Dana & Company, 1856.

Werner, Walter, and Steven Smith. *Wall Street.* New York: Columbia University Press, 1991.

Wills, Garry. *James Madison.* New York: Times Books, 2002.

Wilson, Rufus. *New York: Old and New—Its Story, Streets, and Landmarks.* Philadelphia: J. B. Lippincott, 1902.

Wyld, Lionel. *Low Bridge! Folklore and the Erie Canal.* Syracuse: Syracuse University Press, 1962.

Zakaria, Fareed. *The Future of Freedom: Illiberal Democracy at Home and Abroad.* New York: W. W. Norton, 2003.

ARTICLES AND ONLINE RESOURCES

Brunger, Eric, and Lionel Wyld. "The Grand Canal: New York's First Thruway." *Adventures in Western New York History* (Buffalo and Erie County Historical Society) 12 (1964).

Delong, J. Bradford, 2001. "A Historical Perspective on the New Economy," available at www.j-bradford-delong.net.

Engelbrecht-Wiggans, Richard, and Romas Nonnenmacher. "A Theoretical Basis for Nineteenth-Century Changes to the Port of New York Imported Goods Auction." *Explorations in American Economic History* 36 (1999), pp. 232–45.

Garrett, Wilbur. "George Washington's Patowmack Canal: The Waterway That Led to the Constitution." *National Geographic* 171 (1987), no. 6.

Goetzmann, William, Roger Ibbotson, and Liang Peng. "A New Historical Database for the NYSE 1815–1925: Performance and Predictability." Yale International Center for Finance, Working Paper #00-13, 2000. Available at http://ssrn.com/abstract=236982.

Greenwood, Jeremy, and Ananth Seshadri. "The U.S. Demographic Transition." *American Economic Association Papers and Proceedings*, 2002.

Hawthorne, Nathaniel. "The Canal Boat." *New-England Magazine*, no. 9 (December 1835), pp. 398–409. Also available at www.history.rochester.edu/canal/bib.

Hosack, David. *Memoir of De Witt Clinton: With an Appendix, Containing Numerous Documents, Illustrative of the Principal Events of His Life.* New York: J. Seymour, 1829. Available at www.history.rochester.edu/canal/bib/hosack/Contents.html.

Rothbard, Murray. *The Panic of 1819: Reactions and Policies.* Originally published in 1962. Available at www.mises.org/rothbard/panic1819.pdf.

Rutherford, John. *Facts and Observations in Relation to the Origin and Completion of the Erie Canal.* New York: N. B. Holmes, 1825. Available at www.history.rochester.edu/canal/bib/rutherford/fact1825.htm.

Sawyer, John E. "The Social Basis of the American System of Manufacturing." *Journal of Economic History* 14 (1954), no. 4, pp. 361–79.

Stone, William L. *Narrative of the Festivities Observed in Honor of the Grand Erie Canal Uniting the Waters of the Great Western Lakes with the Atlantic Ocean*. Originally published in 1825. Available at www.history.rochester.edu/canal/bib/colden/App18.html; also in David Hosack, *Memoir of De Witt Clinton*, as appendix, note CC.

Sullivan, James, ed. *The History of New York State*. Originally published in 1927. Vol. 2, ch. 6 is available at www.usgennet.org/usa/ny/state/his/bk2/ch6/pt2.html.

Tatum, Sibyl. *Account of a Journey of Sibyl Tatum with Her Parents*. Typescript, 1830. Purchased from an Internet vendor.

Watson, Elkanah. *The Expedition*. Originally published in 1792. Available at www.nysm.nysed.gov/history/three/bat6.html.

Whitford, Noble E. *History of the Canal System of the State of New York Together with Brief Histories of the Canals of the United States and Canada*. Albany: Brandow, 1905. Available at www.history.rochester.edu/canal/bib/whitford/old1906.

ILLUSTRATION CREDITS

Page 29: *Little Falls* by William Rickarby Miller (1818–1893), 1852, watercolor on paper, Albany Institute of History & Art, 1946.69.

Page 107: *View on the Erie Canal* by John William Hill (1812–1879), 1829, watercolor on paper, I. N. Phelps Stokes Collection, Miriam and Ira D. Wallach Division of Art, Prints and Photographs, The New York Public Library, Astor, Lenox and Tilden Foundations.

Page 201: "Canal at Little Falls" detail of a map by R. H. Pease, 1851, courtesy the Historic American Engineering Record, The Library of Congress.

Page 277: Lockport, New York, from *Building the Nation* by Charles Carleton Coffin, 1882. Courtesy the Library of Congress.

Page 323: *Before the Days of Rapid Transit* by Edward Lamson Henry (1841–1919), c. 1900, color photographic process on paper, Albany Institute of History & Art, x1940.600.56.

INDEX